TECHNICIAN'S GUIDE TO
FIBER OPTICS

TECHNICIAN'S GUIDE TO
FIBER OPTICS

Donald J. Sterling, Jr.

DELMAR PUBLISHERS INC.®

DEDICATION
I dedicate this book to Lynne,
for her patience and support while I wrote it.

Delmar Staff
Executive Editor: Mark Huth
Managing Editor: Barbara Christie
Production Editor: Ruth East
Art Coordinator: Tony Canabush
Design Coordinator: John Orozco

For more information address Delmar Publishers Inc.
3 Columbia Circle, Box 15-015
Albany, New York 12212-5015

COPYRIGHT © 1987
BY DELMAR PUBLISHERS INC.

Printed in the United States of America
Published simultaneously in Canada
by Nelson Canada,
A division of The Thomson Corporation

20 19 18 17 16 15 14 13

Library of Congress Cataloging in Publication Data

Sterling, Donald J., 1951–
    Technician's guide to fiber optics.

    Includes index.
    1. Fiber optics. I. Title.
TA1800.S74   1987   621.36'92   86-13490
ISBN 0–8273–2612–2
ISBN 0–8273–2613–0 (instructor's guide)

Contents

Preface

Technician's Guide to Fiber Optics is intended as an introductory general text on fiber optics. It is suitable for technicians and students in technical and trade schools, community colleges, and first- or second-year students at four-year colleges. It is also well suited for use by companies who wish to acquaint employees with this new technology.

The book is intended to be self-contained. Few prerequisites are required, although readers should be familiar with basic electronics and digital concepts. These topics are briefly reviewed but not developed much past the level needed to understand them within the context of fiber optics, as discussed here.

Use of mathematics is kept limited to practical applications most useful to students. Only a basic acquaintance with trigonometry and logarithms is required.

The book is divided into three parts. Part One, Chapters 1 through 4, attempts to put fiber optics into perspective as a transmission medium. It not only describes the advantages of this new technology over copper counterparts, but also it shows the growing importance of electronic communications and explains basic concepts such as bits and bytes, analog and digital, and light.

Part Two, Chapters 5 through 12, describes in detail fibers, sources, detectors, connectors and splices, and couplers. Emphasis is given to the theoretical and practical aspects of fibers and connectors in particular, because these materials are the most different from their electronic counterparts. Understanding these components will be of most value to anyone actually working with fiber optics.

Part Three, Chapters 13 through 16, attempts to show how fiber-optic systems are put together. It includes discussion of links (with emphasis on how to calculate power and rise-time budgets), installation and special fiber-optic hardware, applications, and fiber-optic equipment.

Each chapter contains a summary of important concepts and a set of review questions. A helpful glossary of key terms completes the book.

ACKNOWLEDGMENTS

First, thanks are due to those who reviewed the manuscript and made valuable suggestions for improvements: Elias A. Awad of Wentworth Institute, Carl Stancil of GTE Sylvania Technical School, Peter Vangel of Springfield Technical College, Sam Conn and Tom Gerwattowski of National Education Center, and Bruce Bothwell of Chemeketa Community College.

I would also like to thank those companies who generously supplied information, artwork, and samples for the book. In particular, I would like to acknowledge the help provided by Loretta Coltrane of GTE, Elaine Hicks of Siecor, Nancy Suey and John Sicotte of Corning, Melissa A. Heck of Anritsu America, Ellen Phillips of Lytel, Carl Blesch of AT&T Bell Laboratories, Marguerite G. Shapalis of ITT Electro-Optical Products, Richard Ryan of Buehler, Andrew Guraxani and A. F. Nelson of Hewlett-Packard, and Robert Chow of Fujitsu Microelectronics.

I would like to thank Richard Aure of Canoga Data Systems for permission to quote from Canoga's application note *Fiberoptic Cable Installation Techniques,* which forms a central part of Chapter 14.

Finally, a special note of thanks is due to AMP Incorporated and, especially, to Jim Pletcher of the product communication division and Terry Bowen, Bob Southard, and Bob Stine of the electro-optics division. AMP was also kind enough to provide many of the illustrations used in the book as well as permission to quote liberally from material I wrote while in its employ, especially *Designer's Guide to Fiber Optics.*

NOTICE TO THE READER

Part One

BACKGROUND

Chapter 1

The Communications Revolution

Fiber optics, as discussed in this book, is simply a method of carrying information from one point to another. An optical fiber is a thin strand of glass or plastic that serves as the transmission medium over which the information passes. It thus fills the same basic function as a copper wire carrying a telephone conversation or computer data. But, unlike the copper wire, the fiber carries light instead of electricity. In doing so, it offers many distinct advantages that make it the transmission medium of choice in applications ranging from telephony to computers to automated factories.

The basic fiber-optic system is a link connecting two electronic circuits. Figure 1–1 shows the main parts of a link, which are as follows:

- *Transmitter,* which converts an electrical signal into a light signal. The source, either a light-emitting diode or laser diode, does the actual conversion. The drive circuit changes the electrical signal fed to the transmitter into a form required by the source.
- *Fiber-optic cable,* which is the medium for carrying the light. The cable includes the fiber and its protective covering.
- *Receiver,* which accepts the light and converts it back into an electrical signal. The two basic parts of the receiver are the detector, which converts the light signal back into an electrical signal, and the output circuit, which amplifies and, if necessary, reshapes the electrical signal before passing it on.
- *Connectors,* which connect the fibers to the source, detector, and other fibers.

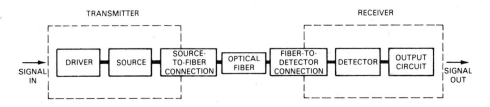

FIGURE 1–1   Basic fiber-optic link (Illustration courtesy of AMP Incorporated)

As with most electronic systems, the transmitter and receiver circuits can be very simple or very complex. These four parts form the essence of the fiber-optic communication alternative. Other components discussed in this book, such as couplers, multiplexers, and distribution hardware, provide the means of forming more complex links and communication networks. But the transmitter, fiber, receiver, and connectors are the basic elements found in every fiber-optic link.

THE HISTORY OF FIBER OPTICS

Using light for communication is not new. Lanterns in Boston's Old North Church sent Paul Revere on his famous ride. Navy signalmen used lamps to communicate by Morse code, and lighthouses for centuries have warned sailors of dangers.

Claude Chappe built an optical telegraph in France during the 1790s. Signalmen in a series of towers stretching from Paris to Lille, a distance of 230 km, relayed signals to one another through movable signal arms. Messages could travel end to end in about 15 min. In the United States, an optical telegraph linked Boston and the nearby island of Martha's Vineyard. These systems were eventually replaced by electrical telegraphs.

The English natural philosopher John Tyndall, in 1870, demonstrated the principle of guiding light through internal reflections. In a demonstration before the Royal Society, he showed that light could be bent around a corner as it traveled in a jet of pouring water. Water flowed through a horizontal spout near the bottom of a container into another container and through a parabolic path through the air. When Tyndall aimed a beam of light out through the spout along with the water, his audience saw the light follow a zigzag path inside the curved path of the water. A similar zigzagging occurs in optical fibers.

A decade later, Alexander Graham Bell patented the photophone (Figure 1–2), which used unguided light to carry speech. A series of lenses and mirrors threw light onto a flat mirror attached to the mouthpiece. The voice vibrated the mirror, thereby modulating the light striking it. The receiver used a selenium detector whose resistance varied with the intensity of light striking it. The voice-

FIGURE 1-2 Alexander Graham Bell's photophone (Illustration courtesy of AMP Incorporated)

modulated sunlight striking the selenium varied the current through the receiver and reproduced the voice. Bell managed to transmit voices over 200 m.

Throughout the early 20th century, scientists made experimental and theoretical investigations into dielectric waveguides, including flexible glass rods. By the 1950s, glass rods were employed for transmitting images, such as a fiberscope, which is used in medicine to look inside the body.

In 1958, Charles Townes and Arthur Schawlow of Bell Laboratories proposed the laser as a source of intense light. By 1960, Theodore Maiman of Hughes Research Laboratory operated the first laser. By 1962, lasing was observed in a semiconductor chip, which is the type of laser used in fiber optics.

The importance of the laser as an important carrier of information was not lost on communication engineers. A laser carrier had an information-carrying capacity four times the magnitude of the radio frequencies then being used. Despite this, the laser was not well suited for open-air line-of-sight transmission. Fog, smog, rain, and other environmental conditions adversely affect the transmission of laser light. It is easier to transmit a laser beam from the earth to the moon than from uptown to downtown Manhattan. The laser, then, was a communications source waiting for a suitable transmission medium.

In 1966, Charles Kao and Charles Hockham, of Standard Telecommunication Laboratory in England, published a paper proposing that optical fibers could be used as a transmission medium if their losses could be reduced to 20 dB/km. They speculated that the current high losses of over 1000 dB/km were the result of impurities in the glass, not of the glass itself. Reducing these impurities would produce low-loss fibers suited for communications.

In 1970, Robert Maurer and colleagues at Corning Glass Works produced the first fiber with losses under 20 dB/km. By 1972, losses were reduced to 4 dB/km in laboratory samples, well below the level Kao and Hockham suggest was required for a practical communication system. Today, losses in the best fibers are around 0.2 dB/km.

Similar advances in semiconductor sources and detectors, connectors, transmission technology, communication theory, and other areas, as well as intense interest in exploiting the possible benefits of fiber optics, led to investigation and trials of fiber-optic systems during the mid- and late 1970s.

The Navy installed a fiber-optic telephone link aboard the USS *Little Rock* in 1973. The Air Force replaced the wiring harness of an A-7 aircraft with an optical link in its Airborne Light Optical Fiber Technology (ALOFT) program in 1976. The wiring harness contained 302 cables with a total length of 1260 m and a weight of 40 kg. The optical replacement had 12 fibers, a total length of 76 m and a weight of 1.7 kg. The military was also responsible for one of the earliest operational fiber-optic data links in 1977, a 2-km, 20-Mbps system connecting a satellite earth station and a data-processing center.

In 1977, AT&T and GTE installed fiber-optic telephone systems carrying commercial telephone traffic. Encouragingly, the systems exceeded the rigid performance standards for reliability imposed by the telephone companies on their equipment and its operation. As a result of successful early trials, optical systems were installed by telecommunications companies throughout the late 1970s and early 1980s. In 1980, AT&T announced plans for an ambitious project for constructing a fiber-optic system running from Boston to Richmond. This system demonstrated the acceptance of the technology for normal, high-capacity systems rather than specialized applications. It showed that fiber optics would be the preferred technology in the future. Fiber optics had clearly arrived as a viable technology.

During the same period, technology also advanced rapidly as the industry matured and gained experience. As late as 1983, advanced single-mode fibers (which will be described in Chapter 4) were felt to be so difficult to use that they would find only specialized application for many years. By 1985, major long-distance carriers such as AT&T and MCI had not only installed single-mode systems but also had chosen them as the standard fiber for future installations.

Table 1–1 lists the first 100 installations using cables supplied by the Siecor Corporation. Notice that the frequency of installations increases each year. During the late 1970s and early 1980s, almost every new fiber-optic installation was a noteworthy event. Today, tracking new systems is nearly impossible as applications proliferate. New installations are becoming routine, and only those that break technological barriers become noteworthy.

Although industries such as computers, information networks, and industrial controls did not embrace fiber optics as quickly as the military and the telecommunications industry, they nonetheless carefully investigated the technology, experimented with it, and closely observed the experiences of other users. As the information age spread the need for better telecommunication systems, it also fostered the growth of fiber-optic applications. Today, fiber optics finds wide use outside of telecommunications.

THE INFORMATION AGE

America, it has been said, is entering the information age, which also means the electronic age. The information age uses electronics to perform the four functions of information processing.

1. Gathering information
2. Storing information
3. Manipulating and analyzing information
4. Moving information from one place to another

The tools of the age are fairly new: computers, electronic offices, a complex telephone network, satellites, information networks and databases, cable television, and so forth. If you look, the evidence of the age surrounds you. The information-services industry has been growing at a rate of 15% a year and shows no signs of slowing down.

In short, America—and much of the rest of the world—is quickly becoming a huge network wired together by electronics. As easily as you can telephone from

PROJECT NUMBER	INSTALLATION	LOCATION	FIBER COUNT*	FIBER TYPE†	COMMENTS
1978					
1	General Telephone Co. of Indiana	Fort Wayne, IN	Low	FWF	Field trial with filled and unfilled buffer tubes. Mechanical and fusion splices used in system.
2	United Telephone Co. of Pennsylvania	Carlisle, PA	Low	FWF	First kevlar cable reinforcing for plow-in installation, filled buffer tubes for water exclusion, dielectric central member. Included aerial, duct, and buried designs.
3	Cablecóm General	Joplin, MO	Low	FWF	First filled core cable. 3 km continuous aerial CATV cables. First hybrid cables for the transmission of electricity and light. First use of an Optical Time Domain Reflectometer in the field.
1979					
4	Western Tele-Communications, Inc.	Colorado Springs, CO	Low	FWF	
5	Western Union Telegraph Co.	New York, NY	Low	ESR	First cable installed with 0.8 to 1.3 μ (extended) spectral response for non-obsolescence and upgradeability. Helical steel tape armoring to withstand 70°C steam bursts.
6	General Telephone Co. of Florida	Tampa, FL	Low	FWF	First armored underwater cable.
7	Mountain States Telephone and Telegraph Co.	Phoenix, AZ	Low	FWF	First 3 km continuous duct cable made and installed.
8	Hawaiian Telephone Co.	Alakea-Kalihi, HI	Low and Medium	FWF	
9	United States Department of Energy	Mercury, NV	Low	FWF	Gas blocked cable for underground nuclear testing.
10	Continental Telephone Co. of Virginia	Woodbridge, VA	Low	FWF	
1980					
11	General Telephone Co. of Pennsylvania	Oil City, PA	Low	FWF	First fully color coded buffer tubes.
12	Hawaiian Telephone Co.	Alakea-Punahua, HI	Low and Medium	FWF	
13	Southwestern Bell Telephone Co.	Greenville, TX	Low	FWF	Microwave entrance link.
14	General Telephone of California	Beaumont, CA	Low	FWF	
15	Mountain States Telephone and Telegraph Company	Colorado Springs, CO	Low	FWF	Longest (30.3 km) repeatered installation of its time in the US.
16	Niagara Mohawk Power Corp.	Syracuse, NY	Low	ESR	First installation within a power company.
17	Continental Telephone Co. of the South	Thomaston, GA	Low	FWF	
18	Continental Telephone Co. of Arkansas	Russellville, AR	Low	FWF	
1981					
19	General Telephone Co. of Florida	Sarasota, FL	Low	FWF	
20	Continental Telephone Co. of California	San Bernadino, CA	Low	FWF	
21	Continental Telephone Co. of California	Taft, CA	Low	FWF	Preconnectorized cable.
22	General Telephone Co. of the Southeast	Ozark, AL	Low	FWF	Digital microwave entrance link.
23	Continental Telephone Co. of Missouri	Wentzville, MO	Low	ESR	

* Fiber count: Low = 16 fibers or less. Medium = 18-40. High = 42 and over.
† First Window Fiber (FWF) operates at 850 nm. Second Window Fiber (SWF) is optimized for operation at 1300 nm, and Extended Spectral Response (ESR) fiber operates at both wavelengths. Single mode (SM) fiber is optimized for operation at 1300 nm.

TABLE 1-1 Early Fiber-optic installations (Courtesy of Siecor Corporation)

PROJECT NUMBER	INSTALLATION	LOCATION	FIBER COUNT*	FIBER TYPE†	COMMENTS
24	Continental Telephone Co. of Virginia	Triangle, VA	Low	ESR	First commercial long wavelength installation. 17.2 km unrepeatered span at 45 Mb/s.
25	Telephone Utilities	Cheney, WA	Low	ESR	
26	United Telephone Co. of Pennsylvania	Harrisburg, PA	Low	FWF	
27	Telephone Utilities	Kalispell, MT	Low	ESR	
28	New England Telephone and Telegraph Co.	White River Junction, VT	Low	FWF	
29	Southwestern Bell Telephone Co.	Conway, AR	Low	FWF	
30	General Telephone Co. of Indiana	Fort Wayne, IN	Low	FWF	Extension of previous project.
1982					
31	General Telephone Co. of Florida	Clearwater, FL	Low	ESR	
32	New York Telephone Co.	Saratoga, NY	Low	FWF	Multiple long aerial spans designed for operation to − 40°C.
33	United States Army	Fort Monmouth, NJ	Low	ESR	
34	Pacific Northwest Bell Telephone Co.	Seattle. WA	Medium and High	ESR	First Mini-Bundle™ cable installation. First 3.5 km continuous high fiber count underwater cable installed.
35	Port Authority of NY	World Trade Center	Low	FWF	Intra-building installation. Long vertical runs.
36	InteCom	University of Chicago	Low	FWF	
37	Hawaiian Telephone Co.	Alakea-Kalihi, HI Alakea-Punahua, HI	Medium	ESR	Extensions of previous installations.
38	Anchorage Telephone Utilities	Anchorage, AK	Low	ESR	
39	Hawaiian Telephone Co.	Kalihi, HI	Low and Medium	ESR	Longest commercial unrepeatered link (32.4 km) at 45 Mb/s, with upgradeability to 90 Mb/s. Up to 36 fibers in single loose tube (three-layer design).
40	Ground Launched Cruise Missile Program	Minneapolis, MN	Low	FWF	Cable designed to resist radiation and severe thermal blasts. Prime contractor is Sperry.
41	New England Telephone and Telegraph Co.	Manchester, NH	Low	FWF	Longest (14 km) unrepeatered short wavelength system (90 Mb/s).
42	United States Air Force	Griffiss Air Force Base	Low	FWF	Intra-building installation.
43	Southwestern Bell Telephone Co.	South Padre Island, TX	Medium	FWF	Installed across causeway between mainland and Padre Island. First subscriber feeder installation.
44	Jet Propulsion Labs	Pasadena, CA	Low	SM	First single-mode installation; evaluation program.
45	Continental Telephone Co. of Indiana	Seymour, IN	Low	SWF	
46	New York Telephone Co.	Long Island, NY	High	FWF	Phase I of metropolitan trunk system. Seven spans, Demonstrated 90 Mb/s over 62.2 km.
			Medium	SWF	Phase II. Five spans, 90 Mb/s over 51.6 km.
47	Fujitsu	Japan	Low	SM	Demonstration of 405 Mb/s system operation over 30 + km single mode cable.
48	AT & T Long Lines	Troy Hill, PA	Low	FWF	
49	MCI Telecommunications, Inc.	New York, NY	Low	SWF	135 Mb/s system.

*Fiber count: Low = 16 fibers or less. Medium = 18-40. High = 42 and over.
†First Window Fiber (FWF) operates at 850 nm. Second Window Fiber (SWF) is optimized for operation at 1300 nm, and Extended Spectral Response (ESR) fiber operates at both wavelengths. Single mode (SM) fiber is optimized for operation at 1300 nm.

TABLE 1-1 (Continued)

PROJECT NUMBER	INSTALLATION	LOCATION	FIBER COUNT*	FIBER TYPE†	COMMENTS
50	Central Trust of China	Taiwan	Low	ESR	Hybrid design: loose tube cable includes fiber and copper wires.
51	Citibank Corp.	New York, NY	Medium	ESR	
52	Hawaiian Telephone Co.	Mililani-Wahiawa, HI	Low	ESR	
53	Mountain States Telephone and Telegraph Co.	Denver, CO	High	ESR/ SWF	Denver Metro project. Two-phase, eight-span project over 90 km long, operating at 90 Mb/s.
54	General Telephone of California	Topanga Canyon, CA	High	SWF	First bandwidth field measurements at 1200 and 1300 nm operation performed on this 30.7 km system.
55	Rochester Telephone Co.	Dansville, NY	Low	ESR	
56	Hawaiian Telephone Co.	Pearl City - Waipahu, HI	Low	ESR	
57	Pacific Northwest Bell Telephone Co./Continental Telephone Co. of the Northwest	Bellingham, WA	Low	SWF	22 km unrepeatered 90 Mb/s system.
58	Rochester Telephone Co.	Perinton, NY	Low	ESR	Microwave entrance link.
59	Hawaiian Telephone Co.	Makila-Lahawa, HI	Low	ESR	
60	New York Telephone Co.	Buffalo, NY	Low	ESR	
61	Southwestern Bell Telephone Co./ Central Telephone Co. of Texas	Humble, TX	Low	ESR	
62	Rochester Telephone Co.	Brighton, NY	Low	ESR	
63	New York Telephone Co.	Marcy, NY	Low	FWF	
1983					
64	General Telephone Co. of Florida	Port Richey, FL	Low and Medium	SWF	
65	General Telephone Co. of Florida	Tampa, FL	Low and Medium	FWF	Includes 200 m of river crossing cable.
66	Continental Telephone Co. of Virginia	Various Routes	Low	SWF	Five long routes, mainly aerial and plow-in cable.
67	Pacific Northwest Bell Telephone Co.	Milwaukie, OR	Medium	ESR	
68	General Telephone Co. of the Southeast	Lexington, KY	Low	SWF	22.5 km unrepeatered span.
69	New York Telephone Co.	Long Island, NY	Medium and High	FWF/ SWF	Phase III of previous projects. Three spans; 90 Mb/s.
70	General Telephone Co. of Wisconsin	Eagle River, WI	Low	SWF	
71	New York Telephone Co.	Albany, NY	Low and Medium	SWF	
72	Southern New England Telephone Co.	Hartford, CT	Medium	SWF	Long (139.4 km) multispan backbone route.
73	Southern Bell Telephone and Telegraph Co./Carolina Telephone and Telegraph Co.	Raleigh, NC	Low	SWF	
74	Central Telephone Co.	Eden, NC	Low	FWF	Interplant local area network.
75	C & P Telephone Co. of Virginia/ Central Telephone Co. of Virginia	Roanoke, VA	Low	SWF	
76	Western Union Telegraph Co.	New York, NY	High	SWF	Teleport project; designed for operation at 135 Mb/s. Two spans.
77	Continental Telephone Co. of the South	Grand Ridge, FL	Low	FWF	6 Mb/s operation. Subscriber feeder loop.
78	United Telephone Co. of Pennsylvania	Chambersburg, PA	Low	FWF	

* Fiber count: Low = 16 fibers or less, Medium = 18-40, High = 42 and over
† First Window Fiber (FWF) operates at 850 nm. Second Window Fiber (SWF) is optimized for operation at 1300 nm, and Extended Spectral Response (ESR) fiber operates at both wavelengths.
  Single mode (SM) fiber is optimized for operation at 1300 nm.

TABLE 1-1 (Continued)

PROJECT NUMBER	INSTALLATION	LOCATION	FIBER COUNT*	FIBER TYPE†	COMMENTS
79	New York Telephone Co.	Long Island, NY	Medium and High	FWF/ SWF	Northern route expansion of previous projects. Five spans; transmission at 90 Mb/s.
80	Southern New England Telephone Co.	Hartford, CT	Low	FWF	Subscriber feeder cable.
81	New England Telephone and Telegraph Co.	White River Junction, VT	Low and Medium	ESR	
82	New York Telephone Co.	Syracuse, NY	Low and Medium	SWF	
83	Alascom, Inc.	Fairbanks, AK	Low	ESR	Cable designed to meet −50°C temperature requirement.
84	Continental Telephone Co. of Missouri	Wentzville, MO	Low	FWF	
85	New York Telephone Co.	Staten Island, NY	Low	SWF	
86	Continental Telephone Co. of Missouri	Warrenton, MO	Low	FWF/ SWF	
87	Continental Telephone Co. of Indiana	Petersburg, IN	Low	SWF	
88	United Telephone Co. of Ohio	Marysville, OH	Low	ESR	Microwave entrance link.
89	Continental Telephone Co. of Indiana	Jasper, IN	Low	SWF	
90	Indiana Bell Telephone Co.	Indianapolis, IN	Low	SWF	
91	Western Union Telegraph Co.	Chicago, IL	Low	SWF	
92	Continental Telephone Co. of Arkansas	Russellville, AR	Low	FWF/ SWF	
93	United Telephone Co. of Ohio	Bellefontaine, OH	Low	ESR	
94	General Telephone Co. of the Southeast	Sumter, SC	Low	SWF	
95	Continental Telephone Co. of the South	Gray, GA	Low	FWF/ SWF	
96	Northern States Power	Red Wing, MN Monticello, MN	Low and Medium	FWF	Nuclear power plant installation.
97	Western Carolina Telephone Co.	Burningtown, NC	Low	SWF	
98	General Telephone Co. of Indiana	Logansport, IN	Low	SWF	
99	Chickasaw Telephone Co.	Sulphur, OK	Low	SWF	
100	Continental Telephone Co. of Indiana/Indiana Bell Telephone Co.	Corydon, IN	Low	SWF	
101	Continental Telephone Co. of New York	Sidney, NY	Low	SM	Expected to be the longest commercial unrepeatered fiber optic link (36.8 km) to date.
102	MCI Telecommunications Corp.	Wilmington, DE Philadelphia, PA	High	SM/ SWF	First phase to be installed in 1983. Part of the largest single mode fiber optic contract awarded to date. Composite single mode/multimode Mini-Bundle™ cable. System designed for operation at 405 Mb/s.
103	MCI Telecommunications Corp.	Los Angeles, CA	High	SM	Includes one copper pair. Cables delivered and installed July, 1983.

*Fiber count: Low = 16 fibers or less. Medium = 18-40. High = 42 and over.
†First Window Fiber (FWF) operates at 850 nm. Second Window Fiber (SWF) is optimized for operation at 1300 nm, and Extended Spectral Response (ESR) fiber operates at both wavelengths. Single mode (SM) fiber is optimized for operation at 1300 nm.

TABLE 1-1 (Continued)

New York to San Francisco, you can send computer data across the country. In offices, large and small computers and other electronic equipment are connected into networks. Engineers use computers at work to design new computers. Home and personal computers abound. This book was written on a computer. It is even possible to write, produce, and publish a book without ever touching a piece of paper. All the information in the book is moved electronically from the writer's word processor to the printer's typesetting and printing equipment.

The following facts give some perspective to the importance of electronics in modern life.

- In 1981, there were an estimated 158 million telephones in the United States, compared to only 39 million in 1950. In addition, the services provided by your telephone company are much more sophisticated.
- From 1950 to 1981, the number of miles of wire in the telephone system increased from 147 million to 1.1 billion.
- In 1985, an estimated 3.7 million personal computers were sold to businesses and consumers in the United States. The personal computers did not exist in 1976; today, they are rapidly becoming commonplace and essential office equipment.
- More than 1500 computer databases are available from which you can obtain information, using a personal computer and the public telephone network.
- AT&T plans to increase its fiber-optic lines from 1300 miles in 1985 to over 11,000 miles in 1987. MCI plans to double its network by adding 18,000 miles by 1987.
- The first fiber-optic telephone system, installed in 1977, operated at 44.7 million bits per second and was capable of carrying 672 voices. By 1985, AT&T decided that future long-distance fiber systems would operate at 1.7 billion bits per second (Gbps)—a capacity of 24,192 voices over a single fiber.

The point is that all these represent sources of information and the means of tying them together. The information may be a telephone call to a friend or the design of a new widget. Important to the growing amount of information is the means of moving it from one place to another. This movement may be a telephone call across the country or from one office to another office a couple of doors away. Telephone companies increasingly use the same digital techniques to send your voice and computer data. Surprisingly, your voice becomes more like computer data: It is converted into digital pulses—numbers—just like computer data. By doing so, the telephone company transmits your telephone call with greater fidelity and reliability. Digital transmission systems—rather than traditional analog systems—are used in most new telephone systems. One study estimates that 34% of telephone central offices used digital transmission equipment in 1984. By 1994, this percentage will increase to 82%. Fiber optics is exceptionally suited to the requirements of digital telecommunications. Coupled with the need to move information is the need to move it faster, more efficiently, more reliably, and less expensively. Fiber optics fills these needs.

THE WIRING OF AMERICA

The oldest and most obvious examples of the wiring of the United States is, of course, the telephone network. At one time, the network carried voices almost exclusively, whereas it now carries voice, computer data, electronic mail, and video pictures.

Cable television is another example. It offers the possibility of interactive television. The cable will not only bring you television, but it will allow you to respond to opinion polls, shop at home, make restaurant and airline reservations, and do many other things. In short, interactive cable TV will become two-way communication. Rather than simply providing television broadcasts to your home, it will allow you to send information from your home. Such systems have been tested experimentally in the United States, Europe, and Japan, and some experiments used fiber optics.

Home banking and financial services, offered in some cities by major banks, are also made possible by the wiring of America. From a personal computer or terminal in your home, you can check account balances, transfer funds between accounts, pay bills, buy stocks and bonds, add to money market accounts, and so forth.

Another area of wiring where computers have invaded is the office. Word processors have replaced secretaries' typewriters. Computers have replaced the accountant's balance books, the manager's spreadsheet, and the engineer's drafting table. Furthermore, a need exists to allow electronic devices to communicate with each other and with the world outside the office. Local area networks fill this need by allowing electronic equipment to be tied together electronically. In addition, many networks also connect to the public telephone network. Again, fiber optics has a role to play in the modern office by providing an efficient transmission medium for networked equipment.

Factories, too, now use electronics extensively. Computers control and monitor activities, and robots perform monotonous, dangerous, or tedious tasks. In advanced factories, various activities are tied together electronically so that all operate harmoniously. But the electrical noise caused by the high-powered machinery can interfere with the control signals. Switching of electrical loads and operating equipment such as arc welders or drills produces electrical noise transients. Because of this noise, a great deal of care, such as shielding cables or separating them from power line sources of noise, must be taken to protect the signals on control lines. Because fibers are immune to electrical noise, fiber optics is a good candidate for the factory.

TELECOMMUNICATIONS AND THE COMPUTER

Until a few years ago, a clear distinction existed between what was part of the telephone system and what was a computer system. Telephone companies, for example, were legally forbidden from competing in the computer marketplace. Today, the distinction, if it still exists, has blurred considerably. Computers send data over telephone lines. Telephone systems turn your voice into computerlike data before transmitting it. Telephone and computer companies now often compete in the same market.

The changes causing this blurring are understandable. Advances in electronic technology meant that each industry used each other's technology. After the breakup of AT&T, the world's largest corporation, in 1982, the distinction between telephony and computers dissolved further. The importance to fiber optics is this: The entire electronic network is becoming a single system. It becomes increasingly difficult to say that this part of the system is the telephone company's responsibility, this part is the computer company's, and this part is the building owner's.

One measure of advances brought by the information age, the wiring of America, and the melding of telephone and computers is the services provided by telephone companies. The service most of us have or had—the ability to dial and receive calls—is called POTS (plain old telephone service). Today, many new services, such as call forwarding and automatic callback, are the beginning of what some call PANS (pretty amazing new services). The telephone companies are aiming toward an integrated-services digital network (ISDN) in which voice, video, and data can be exchanged on the telephone network. Anybody can exchange anything with anybody, anywhere, anytime.

THE FIBER-OPTIC ALTERNATIVE

A wired network such as we have broadly discussed in this chapter needs an efficient medium over which to send information. Traditional technology, such as copper cable and microwave transmission, has limitations that are overcome by fiber optics. Copper cables, for example, have limited capacity and are affected by noise. Microwave transmission, although it has very high capacity, is expensive because of the cost of transmission stations and is limited to line-of-sight transmission, which places severe restrictions on possible locations of transmission equipment. Fiber optics offers capacities well beyond those of copper cables and without the high cost and restrictions of microwave facilities. It will even challenge satellites in some applications. Fiber optics is a new technology whose capabilities are only beginning to be realized. Even though fiber optics possesses many advantages over copper cable, remember that fiber optics is in its infancy, whereas copper cable is a mature technology with much less potential for improvement. Fiber optics promises to be an integral part of the information revolution and the networking of the world.

Fiber optics will affect everyone's life whether it is noticed or not. For example, if you watched the 1980 Winter Olympics from Lake Placid, you may not have realized that fiber optics was used by the broadcasting companies for video and audio links. Here are some examples of areas where fiber optics may touch unnoticed on your life: carrying your voice across the country, bringing television to your home by cable, connecting electronic equipment in your office with

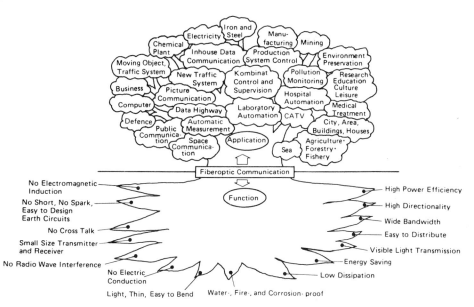

FIGURE 1-3   Applications of fiber optics (Illustration courtesy of Hitachi America Ltd., 1984)

equipment in other offices, connecting electronic subsystems in your car, monitoring operations in a factory.

Figure 1-3 shows the applications and functions of fiber-optic communication as a tree. Branched in the figure are the main application areas, showing that leaves are part of a great, billowing tree. You can clearly see that fiber optics has uses in many areas. The functions, or reasons for using fiber optics, are the roots feeding the applications.

Although fiber optics is a new technology, it has a bright future. It has proven itself in many trials and become the transmission medium of choice in many applications. Its advantages also make it a medium of choice in many other applications in the future.

SUMMARY

- The four parts of a fiber-optic link are the transmitter, cable, receiver, and connectors.
- Modern electronics is an important part of the information age.
- Using electronics to gather, store, manipulate, and move information has resulted in the networking of this country.
- The practical use of fiber optics for communication began in the mid- and late 1970s with test trials. Today it is an established technology.
- The networking will continue.

- Fiber optics has an important role to play in this networking and the information age.
- The advantages fiber optics offers over traditional, older transmission media make it attractive in the information age.

REVIEW QUESTIONS

1. A fiber-optic system transmits information as
   A. Electricity
   B. Light
   C. Sound
   D. Images
2. Name the four main parts of a fiber-optic link.
3. List three of the four functions of information processing.
4. In which of the four areas of information processing is fiber optics most applicable?
5. (True/False) Fiber optics is so specialized a technology that it will never touch most people's lives directly or indirectly.
6. List five areas where fiber optics can be applied. Be as specific as possible.
7. In the 1970s, the greatest interest in and application of fiber optics came from which two of the following:

   A. Military/government   E. Automotive industry
   B. Computer industry   F. Robotics industry
   C. Avionics industry   G. Medical industry
   D. Telephone industry   H. Power industry
8. (True/False) Light as a communication method awaited the invention of the laser and the optical fiber.

Chapter 2

Information Transmission

The last chapter discussed the importance of electronic information to our modern world. The purpose of this chapter is to introduce important aspects of signals and their transmission. An understanding of the underlying principles that allow modern electronic communication is fundamental to an understanding and appreciation of fiber optics. The ideas presented here are fundamental not only to fiber optics but to all electronic communication. We do not propose to offer an in-depth survey of communication theory and its application to electronics. The purpose of this chapter is to introduce terms and principles without which any discussion of fiber optics would be stymied.

COMMUNICATION

Communication is the process of establishing a link between two points and passing information between them. The information is transmitted in the form of a signal. In electronics, a signal can be anything from the pulses running through a digital computer to the modulated radio waves of an FM radio broadcast. Such passing of information involves three activities: encoding, transmission, and decoding.

Encoding is the process of placing the information on a carrier. The vibration of your vocal chords places the code of your voice on air. The air is the carrier, changed to carry information by your vocal chords. Until it is changed in some way, the carrier carries no information. It is not yet a signal. A steadily oscillating electronic frequency can be transmitted from one point to another, but it also carries no intelligence unless information is encoded on it in some way. It is merely the carrier onto which the information is placed. Conveying information, then, is the act of modifying the carrier. This modification is called *modulation*.

Figure 2-1 shows the creation of a signal by impressing information on a carrier. A high-frequency carrier, which in itself carries no information, has impressed on it a lower-frequency signal. The shape of the carrier is now modulated by the information. Although the simple example in the figure conveys very little information, the concept can be extended to convey a great deal of informa-

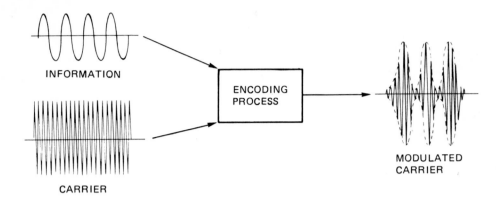

INFORMATION

CARRIER

ENCODING
PROCESS

MODULATED
CARRIER

FIGURE 2-1 A signal

tion. A Morse-code system can be based on the example shown. On the unmodulated carrier can be impressed a low frequency for one of two durations corresponding to the dots and dashes of the code.

Once information has been encoded by modulating a carrier, it is *transmitted*. Transmission can be over air, on copper cables, through space to a satellite and back, or through optical fibers.

At the other end of the transmission, the receiver separates the information from the carrier in the *decoding* process. A person's ear separates the vibrations of the air and turns them into nerve signals. A radio receiver separates the information from the carrier. For the encoded signal in Figure 2-1, the receiver would strip away and discard the high-frequency carrier while keeping the low-frequency signal for further processing. In fiber optics, light is the carrier on which information is impressed.

Assume you are writing a letter on a personal computer to a friend. The letters you press at the keyboard are encoded into the digital pulses the computer requires. Once finished, you wish to send the letter to your friend's computer. You will need a *modem*, a device that reencodes the digital pulses into audio pulses easily sent over the telephone lines to the central telephone office.

The central office receives your transmission. It reencodes your signal into digital pulses and transmits it with many other signals over an optical fiber to another central office. This office decodes the digital pulses to audio pulses. It sends this information to your friend, whose modem further decodes the signal into computer pulses. Your friend can read your letter by displaying it on the computer screen or by printing it out.

Many ways exist to modulate the carrier. Figure 2-2 shows three common ways.

1. *Amplitude modulation (AM).* Amplitude modulation is used in AM radio. Here the amplitude of the carrier wave is varied to correspond to the amplitude of the information.

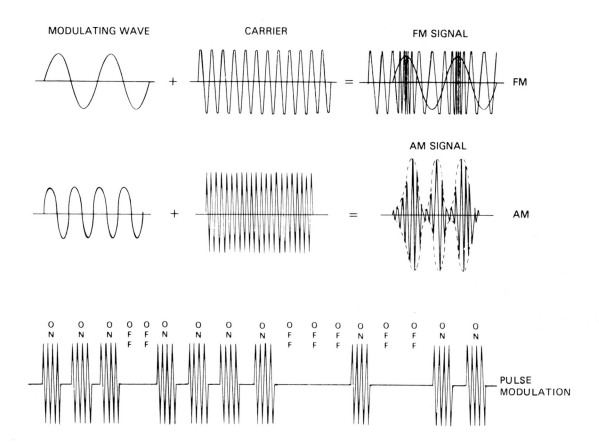

FIGURE 2-2   Examples of modulation (Illustration courtesy of AMP Incorporated)

2. *Frequency modulation (FM).* Frequency modulation changes the frequency of the carrier to correspond to differences in the signal amplitude. The signal changes the frequency of the carrier, rather than its amplitude. FM radio uses this technique.
3. *Pulse-coded modulation (PCM).* Pulse-coded modulation converts an analog signal, such as your voice, into digital pulses. Your voice can be represented as a series of numbers, the value of each number corresponding to the amplitude of your voice at a given instance. PCM is the main way that voices are sent over a fiber-optic telephone system. We will look more closely at it in a moment.

ANALOG AND DIGITAL

We live in an analog world. "Analog" implies continuous variation, like the sweep of the second hand around the face of an electric watch. Your voice is

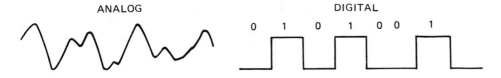

FIGURE 2-3 Analog and digital information

analog in that the vibrations that form sound can vary to any point within your range of vocal frequencies. Sound is analog, in that it is continuously varying within a given range. Electronic circuits, such as transmitters and receivers, use analog circuits to deal with these continuous variations. Indeed, before the advent of computers, most electronic engineering dealt with analog circuits.

Consider a light operated by a dimmer. Turning the dimmer to adjust the light's brightness is an analog operation. The brightness varies continuously. There are no discrete levels, so that you can easily turn the light a little brighter or a little less brighter.

In contrast, "digital" implies numbers—distinct units like the display of a digital watch. In a digital system, all information ultimately exists in numerical values of digital pulses.

In contrast to the dimmer light, a three-way lamp is digital. Each setting brings a specific level of brightness. *No levels exist in between.* Figure 2-3 shows analog and digital signals.

Important to electronic communication is that analog information can be converted to digital information, and digital information can be turned into analog information. You may be familiar with digital stereo systems. In these systems, music is encoded into digital form as a series of numbers that represent the analog variations in the music. The electronics in the playback system must reconvert the digital code into analog music.

Digital Basics: Bits and Bytes

The basis of any digital system is the *bit* (short for *binary digit*). The bit is the fundamental unit of digital information, and it has only one of two values: 1 or 0. Many ways exist to represent a bit. In electronics, the presence or absence of a voltage level is most common: One voltage level means a "1"; a second level means a "0." Unfortunately, the single bit 1 or 0 can only represent a single state, such as "on" or "off." For example, a lamp can be represented by 0 when turned off and by 1 when turned on:

Off = 0
On = 1

A single bit of information, then, is of very limited usefulness. But we can describe the state of the three-way lamp with 2 bits:

Off = 00
On = 01
Brighter = 10
Brightest = 11

Two bits allow us to communicate more information than we can with a single bit. In our lamp example, 2 bits allowed us to distinguish between four distinct states. The more bits we use in a unit, the more potential information we can express. A digital computer typically works with units of 8 bits (or multiples of 8, such as 16 and 32).

An 8-bit group is called a *byte*. A single byte allows all the letters, numbers, and other characters of a typewriter or computer keyboard to be encoded by a number, with room left over. Eight bits permits 256 different meanings to be given to a pattern of 1s and 0s. The number of different combinations or meanings for any given length of bits n is equal to 2 to the nth power, 2^n. For example, 16 bits yields 65,536 combinations, since $2^{16} = 65,536$. Each time a bit is added, the number of possible different values doubles.

A pulse train is often shown in its ideal form, as the string in Figure 2-4. The pulses go from one state to another instantaneously. Such diagrams show ideal pulses and the main essentials of the pulse train, so engineers and technicians can compare trains.

A pulse train represents the 1s and 0s of digital information. The train can depict high- and low-voltage levels or the presence and absence of a voltage. In electronics, a digital 1 can be a voltage pulse or a higher voltage level. A digital 0 can be the absence of the pulse or a lower voltage level. Thus we can speak of a 1 with terms such as "on" or "high," and of 0 with terms such as "off" or "low."

A real pulse does not occur instantaneously as it is shown doing in Figure 2-4. An electronic circuit has a finite response time, so it takes time for a voltage or optical pulse to turn on and off or switch between high and low levels. The pulse must also stay on for a brief time. Still, in a computer system, turning a pulse on and off can require only millionths or billionths of a second, to allow thousands or millions of pulses per second.

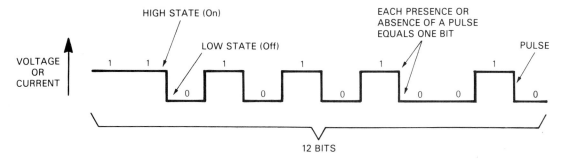

FIGURE 2-4   Ideal pulse train (Illustration courtesy of AMP Incorporated)

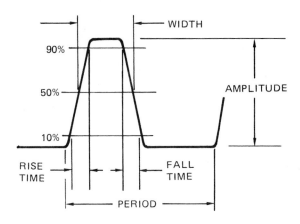

FIGURE 2-5 Parts of a pulse

Engineers dealing with digital systems must consider the shape of the pulse. Figure 2-5 defines the parts of the pulse.

- *Amplitude* is the height of the pulse. It defines the level of energy in the pulse. The energy can be voltage in a digital system or optical power in a fiber-optic system. Notice that different types of energy are used for different types of systems.
- *Rise time* is the time required for the pulse to turn on—to go from 10% to 90% of its maximum amplitude.
- *Fall time,* the opposite of rise time, is the time required to turn off, measured from 90% to 10% of amplitude. Rise time and fall time may or may not be equal.
- *Pulse width* is the width, expressed in time, of the pulse at 50% of the amplitude.
- *Bit period* is the time given for the pulse. Most digital systems are clocked or timed. Pulses must occur in the time allotted by the system for a bit period. For example, suppose a string of five 0s occurs. How do we know it is not four 0s or six 0s? We can detect no changes in the pulse train. We know because nothing has occurred in 5 bit periods of the clock.

A clock is an unchanging pulse train that provides timing by defining bit periods. Each bit period can be defined as one or more clock timing periods. The clock is loosely comparable to a musician's metronome: Both provide the basic timing beat against which operation (pulses or musical notes) is compared.

Rise time is especially important in electronics and fiber optics. Rise time limits the speed of the system. The speed at which pulses can be turned on and off will determine the fastest rate at which pulses can occur. The easiest way to increase the speed of a system is by decreasing pulse rise and fall times, thereby turning pulses on and off faster. Thus more pulses can occur in a given time. Even when pulse amplitude and pulse width remain the same, faster rise times bring faster operating speeds. In other cases, we wish to maintain the pulse width as a certain proportion of pulse time. Faster rise and fall times allow shorter pulse widths, which increase operating speed even more. Conversely, slowing the rise time slows the operating speed.

As we will see in Chapter 10, the bits of 1s and 0s can be formed in other ways than simply turning a voltage on and off. Different formats for coding 1s and 0s offer various advantages and disadvantages in transmitting information. Here we have showed a 1 as a high voltage and 0 as a low voltage occurring in a given bit period. Other codes use both high and low voltages within a bit period to encode a 1 or 0. Still, the essential thing here is that digital systems use 1s and 0s, high and low states, to encode information.

Why Digital?

Computers use digital signals because it is their nature: They are digital by design. Although a computer does not have to use digital techniques and binary numbers, it does so because they are the most convenient way to construct a computer that operates electronically. Transistors and integrated circuits made up of thousands of transistors operate as very fast on/off switches.

Telephone companies use digital techniques and turn your voice into a digital signal before transmitting it. What is the advantage? A signal transmitted any distance becomes distorted. Even if your voice is faithfully reproduced electronically for transmission as an analog signal, it still is distorted by the time it reaches the receiver. Unfortunately, the receiver cannot correct the distortion, because it has no way of knowing what the original signal looked like.

For digital pulses, the situation is reversed. A digital pulse has a defined shape. The receiver knows what the original pulse looked like. A receiver only needs to know information about the pulses: How many pulses? When were they sent? and so on. The receiver can be designed to rebuild distorted pulses into their original shape to faithfully reproduce the original signal.

INFORMATION-CARRYING CAPACITY

Any transmission path carrying signals has limits to the amount of information it can carry. The amount of information that the path can carry is its *information-carrying capacity*. Several ways exist to describe this capacity. In telephony, capacity is expressed in voice channels. A voice channel is the bandwidth or range of frequencies required to carry a single voice. Since the upper end of the human voice is about 4 kHz, a single voice channel must have a bandwidth of 4 kHz. In the early days of telephones, each wire carried only a single voice. Today, a single telephone line carries hundreds or thousands of voices simultaneously. It thus has a capacity of thousands of voice channels.

Bandwidth is proportional to the highest rate at which information can pass through it. If an optical fiber, for example, has a bandwidth of 400 MHz, it can carry frequencies up to this limit.

In digital systems, capacity is given in bits per second (bps) or *baud*. For a telephone system, a single digital voice channel requires 64,000 bps. A digital system, therefore, requires more bandwidth than a comparable analog system. An analog telephone system requires 4 kHz per voice; a digital system requires over 16 times more—64 kHz. Simple digital telephone systems, which carry 672 voices one way over a single line, have an operating speed of 44.7 megabits per second (Mbps).

The methods that allow such systems to be built are PCM and multiplexing.

PCM AND MULTIPLEXING

The technique used to turn your analog voice into a digital signal is *PCM*. Time-division multiplexing is the means that allows several voice channels to be transmitted on a single line.

Communication theory states that an analog signal, such as your voice, can be digitally encoded and decoded if it is sampled at twice the rate of its highest frequency. The high end of the speaking voice in telephony is 4000 Hz, which means that a PCM system must sample the voice 8000 times a second. *Sampling* means that the system looks at the amplitude of the vocal frequencies. Each sample is converted into an 8-bit number. Eight bits allow 256 different amplitudes to be encoded. Since each sample requires 8 bits and the sampling occurs 8000 times a second, the bandwidth required for a single voice channel is 64,000 bps (8000 samples × 8 bits/sample = 64,000 bits). At the receiver end, the same rules used to encode the sample are used to decode the sample and to reconstruct the analog voice signal. Figure 2–6 shows the idea of PCM. There are many ways to design and use PCM systems, but the basic idea is as we have described.

FIGURE 2-6   PCM (Illustration courtesy of AMP Incorproated)

Sending 64,000 bps does not make efficient use of the capacity of a typical transmission path. To make full use of the capacity, telephone companies send several voice channels over the path. The appearance is that all channels are transmitted simultaneously, although they really are not. Part of one voice is first sent, and then part of the second, part of the third, and so on, in an interleaving pattern of channels. The device that permits this combination of different signals onto a single line is a *multiplexer*. The speed of the system easily allows all signals to be sent in turn. A *demultiplexer* at the receiver performs the opposite operation and separates the signals.

Time-division multiplexing (TDM) is the name given to the method of multiplexing that assigns parts of each voice channel to specific time slots. Other forms of multiplexing, such as *frequency-division multiplexing* (FDM) and *wavelength-division multiplexing* (WDM), exist. FDM assigns different information channels to different carrier frequencies. Cable television uses FDM. WDM, which is unique to optical communications, is discussed in Chapter 11.

Fiber optics is important to such telephone systems because its information-carrying capacity exceeds wire-based systems.

THE DECIBEL

The *decibel* is an important unit that you will use continually in fiber optics as well as in electronics. It is used to express gain or loss in a system or component. A transistor, for example, can amplify a signal, making it stronger by increasing its voltage, current, or power. This increase is called *gain*. Similarly, loss is a decrease in voltage, current, or power. The basic equations for the decibel are

$$dB = 20 \log_{10}\left(\frac{V_1}{V_2}\right)$$

$$dB = 20 \log_{10}\left(\frac{I_1}{I_2}\right)$$

$$dB = 10 \log_{10}\left(\frac{P_1}{P_2}\right)$$

where V is voltage, I is current, and P is power. The decibel, then, is the ratio of two voltages, currents, or powers. Notice that voltage and current are 20 times the logarithmic ratio, and power is 10 times the ratio.

A basic use of the decibel is to compare the power entering a system, circuit, or component to the power leaving it. The decibel value shows how the device affects the circuit. A transistor, for example, usually increases the power. Other components may decrease the power. P_1 is the power out, and P_2 is the power in.

$$dB = 10 \log_{10}\left(\frac{P_{out}}{P_{in}}\right)$$

$$P_{in} \rightarrow \boxed{ckt} \rightarrow P_{out}$$

Another use of the decibel is to describe the effect of placing a component in a system. Sometimes, for example, it is necessary to split a cable and apply connectors so that the two cable halves can be connected and disconnected. Insertion of these connectors causes some loss, which is expressed in decibels. In other cases, devices called *attenuators* are purposely placed in a circuit to provide loss.

In fiber optics, we deal mostly with loss and optical power. (The electronic circuits in the transmitter and receiver may use voltage and current and provide gain.) The source emits optical power. As light travels through the fiber to the receiver, it loses power. This power loss is expressed in decibels. For example, if the source emits 1000 microwatts (μW) of power and the detector receives 20 μW, the loss through the system is about 17 dB:

$$\text{Loss} = 10\ \log_{10}\left(\frac{P_r}{P_{tr}}\right)$$

$$= 10\ \log_{10}\left(\frac{20}{1000}\right)$$

$$= -16.989\ \text{db}$$

where P_{tr} is the power transmitted from the source and P_r is the power received by the receiver.

Table 2-1 shows the amount of power remaining for different decibel values. A 10-dB loss (–10 dB) represents a loss of 90% of the power; only 10% remains. An additional loss of 10 dB increases the loss by an order of magnitude. A useful figure to remember is 3 dB, which represents a loss of one half of the power.

Fiber-optic links easily tolerate losses of 30 dB, meaning that 99.9% of the power from the source is lost before it reaches the detector. If the source emits

Loss (dB)	Power Remaining (%)	Loss (dB)	Power Remaining (%)
0.1	97.7	4	39.8
0.2	95.5	5	31.6
0.3	93.3	6	25.1
0.4	91.2	7	19.9
0.5	89.1	8	15.8
0.6	87.1	9	12.6
0.7	85.1	10	10.0
0.8	83.2	20	1.0
0.9	81.1	30	0.1
1	79.4	40	0.01
2	63.1	50	0.001
3	50.1	60	0.001

TABLE 2-1 Loss and the decibel

1000 μW of power, only 1 μW reaches the detector. Of concern in fiber optics are the proper balance of the source power, the losses in the system, and the sensitivity of the detector to weak light signals.

Remember that a decibel expressing loss is a negative unit. In fiber optics, it is common practice to omit the negative sign and speak of a loss of 6 dB, say. This loss is actually –6 dB. If you solved the equation, the result would be –6 dB. But in talking, and even in data sheets, the negative sign is omitted, with little confusion being caused. But if you have occasion to use a loss figure in an equation, do not forget to make the number negative! (The confusion arises because some equations have been adjusted to accept a loss figure as a positive number.)

Sometimes the ratio for calculating loss or gain uses a constant for P_2. In fiber optics, this value is usually 1 milliwatt (mW).

$$\text{dBm} = 10 \log_{10} \left(\frac{P}{1 \text{ mW}} \right)$$

dBm means "decibels referenced to a milliwatt." In this case, the negative sign is almost always used. A value of –10 dBm means that P is 10 dB less than 1 mW, or 100 μW. Similarly, –3 dBm is 500 μW. Communication engineers and technicians use dBm units extensively.

Figure 2-7 relates various dBm values to milliwatt and microwatt levels. Notice that a small range of the dBm values allows a great range of power values to be expressed.

$$
\begin{aligned}
10 \text{ mW} &= +10 \text{ dBm} \\
5 \text{ mW} &= + 7 \text{ dBm} \\
500 \ \mu\text{W} &= - 3 \text{ dBm} \\
100 \ \mu\text{W} &= -10 \text{ dBm} \\
50 \ \mu\text{W} &= -13 \text{ dBm} \\
10 \ \mu\text{W} &= -20 \text{ dBm} \\
5 \ \mu\text{W} &= -23 \text{ dBm} \\
1 \ \mu\text{W} &= -30 \text{ dBm} \\
100 \text{ nW} &= -40 \text{ dBm} \\
10 \text{ nW} &= -50 \text{ dBm} \\
1 \text{ nW} &= -60 \text{ dBm} \\
100 \text{ pW} &= -70 \text{ dBm} \\
10 \text{ pW} &= -80 \text{ dBm} \\
1 \text{ pW} &= -90 \text{ dBm}
\end{aligned}
$$

FIGURE 2-7   Power to dBm conversion

Another unit sometimes used is dBμ—decibel referenced to 1 μW. This is the same as dBm, except the reference value is 1 μW rather than 1 mW.

$$\text{dB}\mu = 10 \log_{10} \left(\frac{P}{1 \ \mu\text{W}} \right)$$

SUMMARY

- Communication consists of encoding, transmitting, and decoding information.
- Digital systems work with bits representing 1 and 0.
- Information-carrying capacity expresses the amount of information that can be transmitted over a communications path.
- PCM turns an analog signal into a digital signal.
- Multiplexing allows many signals to be sent over a single communications path.
- The decibel is the primary unit for indicating gain or loss in a system.

REVIEW QUESTIONS

1. Name the three activities involved in communication. Give an example of each.
2. Define modulation.
3. Sketch an analog signal. Sketch a digital signal.
4. Is the information 1101101 analog or digital?
5. Sketch a pulse. Label its amplitude, rise time, fall time, and pulse width.
6. What is the capacity of an analog voice channel? Of a digital voice channel?
7. For a PCM system as described in this chapter, how many bits per second are required to transmit five voice channels? 50 voice channels?
8. Name the technique that allows five voice channels to be transmitted simultaneously over a single optical fiber.
9. Which requires greater bandwidth, an analog voice channel or a digital voice channel?
10. How many milliwatts are in a signal having a power of 0 dBm?

Chapter 3

Fiber Optics as a Communications Medium: Its Advantages

In its simplest terms, fiber optics is a communications medium linking two electronic circuits. The fiber-optic link may be between a computer and its peripherals, between two telephone switching offices, or between a machine and its controller in an automated factory. Obvious questions concerning fiber optics are, why go to all the trouble of converting the signal to light and back? Why not just use wire? The answer lies in the following advantages of fiber optics.

- Wide bandwidth
- Low loss
- Electromagnetic immunity
- Security
- Light weight
- Small size
- Safety

The importance of each advantage depends on the application. In some cases, wide bandwidth and low loss are overriding factors. In others, security or safety are the factors that lead to the use of fiber optics. The following pages discuss each advantage in detail.

WIDE BANDWIDTH

The last chapters showed the increasing concern with sending more and more information electronically. Potential information-carrying capacity increases with the bandwidth of the transmission medium and with the frequency of the carrier. From the earliest days of radio, useful frequencies have pushed upward five orders of magnitude, from about 100 kHz to about 10 GHz. The frequencies of light are several orders of magnitude above the highest radio waves. The invention of the laser, which uses light as a carrier, in a single step increased the potential range four more orders of magnitude—to 100,000 GHz (or 100 terahertz [THz]). Optical fibers have a potential useful range to about 1 THz, although this range is far from being exploited today. Still, the practical bandwidth of optical fibers exceeds that of copper cables. Furthermore, the information-carrying possibilities of fiber optics have only begun to be exploited, whereas the same potentials of copper cable are pushing their limits.

As mentioned earlier, telephone companies increasingly use digital transmission. The higher bandwidth of fiber optics allows a higher bit rate and, consequently, more voice channels per cable. For compatibility to exist among all telephone carriers, the rates at which different transmission lines carry information are generally fixed by a system known as the North American Digital Telephone Hierarchy.

Table 3–1 shows the hierarchy for coaxial cable and optical fibers. Notice that the transmission rates are the same at lower capacities, but that fibers offer higher capacities than coaxial cables. Repeater spacings, or the distance a signal can travel before it must be amplified, is discussed in the next section. The figures given are for a single fiber or coaxial conductor. Many installations have several of these bundled into a single cable, which effectively increases capacity by the number of conductors (fiber or copper).

For high-speed fiber systems, bit rates, and especially repeater spacings, are not yet fixed. The technology is advancing so fast that the "high end" has not been stabilized. Some of the numbers are based on announced systems that were either being planned or installed at the time of this writing. For example, a type of fiber known as dispersion-shifted single mode became commercially available. Using this type of fiber in place of a nonshifted fiber allows a repeater spacing to be doubled from 40 to 80 km. Future systems will probably also double the highest rate for fibers from 1700 to 3400 Mbps. Such speeds and repeater spacings are a dramatic improvement over the first fiber-optic telephone systems, which had speeds of 45 Mbps and repeater spacings of 6.4 km.

MEDIUM	BIT RATE (Mbps)	VOICE CHANNELS	REPEATER SPACING (km)
Coaxial	1.5	24	1–2
	3.1	48	
	6.3	96	
	45	672	
	90	1,344	
Fiber	45	672	6–15 (multimode)
	90	1,344	30–40+ (single mode)
	180	2,688	
	405 to 435	6,048	
	565	8,064	
	1,700	24,192	

TABLE 3-1 Digital telephone transmission rates

Similar advances will not occur for coaxial systems: Optical systems will become the favored solution for long-distance, high-data-rate applications.

Another consequence of the high bandwidth of fiber optics is that it permits transmission of channels requiring much greater bandwidth than a voice. Television and teleconferencing require a channel capacity 14 to 100 times that of a digitally encoded voice. The bandwidth of the fiber allows these signals also to be multiplexed through the fiber. Fibers allow telephone companies to simultaneously transmit voice, data, and video, which is important to the information age.

As of late 1985, the record for transmission over a single optical fiber, achieved by AT&T Bell Laboratories, was 20,000 Mbps (20 Gbps) over a distance of 68 km. The output of 10 lasers, each operating at a slightly different wavelength and at a data rate of 2000 Mbps (2 Gbps), was multiplexed on a fiber. Although this record was produced under laboratory conditions, today's laboratory experiments are tomorrow's commercial products.

LOW LOSS

Bandwidth is an effective indication of the rate at which information can be sent. Loss indicates how far the information can be sent. As a signal travels along a transmission path, be it copper or fiber, the signal loses strength. This loss of strength is *attenuation*. In a copper cable, attenuation increases with modulation frequency: the higher the frequency of the information signal, the greater is the loss. In an optical fiber, attenuation is flat; loss is the same at any signaling frequency up until a very high frequency. Thus, the problem of loss becomes greater in a copper cable as information-carrying capacity increases.

Figure 3–1, by showing the loss characteristics for fibers, twisted pairs, and coaxial cable, such as are used in telephone systems, demonstrates the usable ranges for these transmission media. Loss in coaxial cable and twisted-pair wires increases with frequency, whereas loss in the optical cable remains flat over a very wide range of frequencies. The loss at very high frequencies in the optical fiber does not result from additional attenuation of the light itself by the fiber. This attenuation remains the same. The loss is caused by loss of information, not by optical power. Information is contained in the variation of the optical power. At very high frequencies, distortion causes a reduction or loss of this information.

The point is that the effects of loss that must be accounted for in a system depend on the signal frequency. What is suitable for a system working at one speed may not work at another frequency. This need to consider different signaling speeds complicates design. High-frequency design, for example, is more difficult than low-frequency design. One cannot simply increase operating speeds without taking into account the effects of speed on performance. In a fiber-optic system, the loss is the same at all operating speeds within the fiber's specified bandwidth.

Severe attenuation requires repeaters at intermediate points in the transmission path. For copper cables, repeater spacings, in general, decrease as operating speeds increase. For fibers, the opposite is true: Repeater spacings increase along

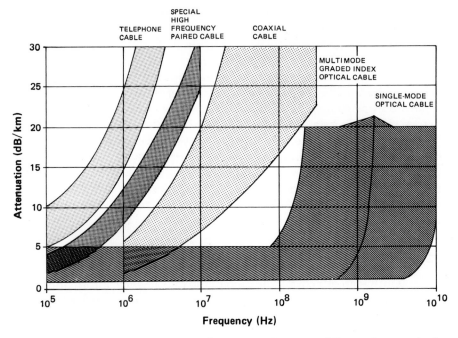

FIGURE 3-1 Attenuation versus frequency (Courtesy of Siecor Corporation)

with operating speeds, because at high data rates very efficient, low-loss fibers are used.

The first transatlantic fiber-optic telephone link, being installed by AT&T and expected to be in operation by 1988, will carry 37,800 simultaneous voice conversations each direction on a pair of fibers. Repeater spacings will be 35 km or more. In contrast, the best transatlantic coaxial system handles 4200 conversations and has a repeater spacing of only 9.4 km. The possibility exists that fiber-optic systems will carry 200 Mbps over 80 to 100 km without a repeater.

The combination of high bandwidth and low loss has made the telephone industry probably the heaviest user of fiber optics. It allows them not only to send more information over a transmission path but also to extend repeater spacings. Repeaters, after all, are electronics systems that are costly to build, install, and maintain. Fewer repeaters means less costly systems.

ELECTROMAGNETIC IMMUNITY

Unlike copper cables, optical fibers do not radiate or pick up electromagnetic radiation. Any copper conductor acts like an antenna, either transmitting or receiving energy. One piece of electronic equipment can emit electromagnetic interference (EMI) that disrupts other equipment. Among reported problems resulting from EMI are the following:

- The military reported such a high concentration of electronic equipment in Vietnam that increases in the population of electronic devices beyond a crucial density made it impossible for the equipment to operate.
- An electronic cash register interfered with aeronautical transmissions at 113 MHz.
- Coin-operated video games interfered with police radio transmissions in the 42-MHz band.
- Some personal computers tested by the Federal Communications Commission (FCC) in 1979 emitted enough radiation to disrupt television reception several hundred feet away.
- Electrostatic discharges (ESD), discharged into computer terminals by operators, have garbled the computer memory, erased work in progress, and even destroyed circuits. (ESD is the shock you get when you walk across a carpet on a dry day and touch a doorknob. Such discharges easily contain 15 to 25 kV.)
- An explosion caused by static electricity killed three workers at Cape Kennedy in the 1960s.
- A manufacturer of gasoline pumps found that CB radio transmissions zeroed remote pump readouts.
- Airport radar erased taxpayers' records in a computer memory bank.

EMI is a form of environmental pollution with consequences ranging from the merely irksome to the deadly serious. As the density of electronic equipment grows (and it will as part of the information age), the potential for EMI problems also increases. To combat such problems, the FCC in 1979, for example, issued stringent regulations limiting EMI from computing devices. Regulatory agencies in Europe have issued similar regulations.

Here is a simple way to demonstrate the effects of EMI if you have access to a personal computer or computer terminal. Place an AM radio near the computer while it is running. Tune through the dial. At some point, the radio should pick up the computer, so that you will be able to listen to the computer operate. Try running different programs to hear differences. What you hear is EMI.

Cables interconnecting equipment can be one of the main sources of EMI. They can also be one of the main receiving antennas carrying EMI into equipment. Cables act just like the radio antenna in the foregoing EMI demonstration.

Since fibers do not radiate or receive electromagnetic energy, they make an ideal transmission medium when EMI is a concern. Some factories use fiber optics because of its immunity. These applications do not require the high bandwidth or low loss of fibers. Equipment such as motors that switch on and off can be a source of EMI that disrupts signal lines controlling the equipment. Using fibers rather than copper cables eliminates the problems.

High-voltage lines can present another problem, since they also emit energy. Copper signal cables cannot be run next to such lines without special precautions, because energy from the high-voltage line couples onto the signal line. Fiber-optic lines can be run beside high-voltage lines with no detrimental effect, since no energy couples onto it from the high-voltage line.

One consequence of the fiber's electromagnetic immunity is that signals do not become distorted by EMI. Digital transmission requires that signals be transmitted without error. EMI can be a cause of error in electrical conductor transmission systems. A burst of EMI may appear as a pulse, where no pulse occurred in the original pulse stream. Fibers offer very high standards in error-free transmission.

SECURITY

One way to eavesdrop is to tap a wire. Another way is to pick up energy radiated from a wire or equipment (a form of EMI). Years ago, the United States discovered a foreign power doing just that to our embassies. A sensitive antenna in a nearby building was secretly picking up energy radiated from electronic equipment in the embassy. The antenna was receiving EMI much like the radio example discussed earlier. This energy, though, included top-secret and classified data. Businesses also spend millions of dollars each year protecting their secrets, such as encrypting data before it is transmitted.

Fiber optics is a highly secure transmission medium. It does not radiate energy that can be received by a nearby antenna, and it is extremely difficult to tap a fiber, and virtually impossible to make the tap undetected. Both government and business consider fiber optics a secure medium.

LIGHT WEIGHT

A glass fiber weighs considerably less than a copper conductor. A fiber-optic cable with the same information-carrying capacity as a copper cable weighs less than the copper cable because the copper requires more lines than the fiber. For example, a typical single-conductor fiber-optic cable weighs 9 lb/1000 ft. A comparable coaxial cable weighs nine times as much—about 80 lb/1000 ft. Weight savings are important in such applications as aircraft and automobiles.

SMALL SIZE

A fiber-optic cable is smaller than its copper counterparts. In addition, a single fiber can often replace several copper conductors. Figure 3–2 shows a comparison of coaxial cable and a fiber-optic cable used in digital telephony. The copper cable is 4.5 in. in diameter and can carry as many as 40,300 two-way conversations over short distances. The fiber-optic cable, which contains 144 fibers in its 0.5-in.-diameter structure, has the capacity to carry 24,192 conversations on each fiber pair or nearly 1.75 million calls on all the fibers. The fiber-optic cable greatly exceeds the capacity of the coaxial cable even though it is almost 10 times smaller.

The small size of fiber-optic cables makes them attractive for applications where space is at a premium:

FIGURE 3-2   Size comparison: coaxial cable and fiber-optic cable (Courtesy of AT & T Bell Laboratories)

- Aircraft and submarines, where the use of every square inch is critical. Not only do the fiber-optic cables use less of the valuable space, they can be placed in areas where copper cables cannot be. In short, they make efficient use of the space.
- Underground telephone conduits, especially in cities. Here, not only are conduits often filled to capacity, but building new conduits to allow expanded services is very costly. Additional space may be unavailable. Fiber-optic cables replace the copper cables, often offering greater capacity in less space. A large copper cable that fills a conduit can be replaced by a smaller fiber cable, with room left for new cables in the future.
- Computer rooms, where the cables between equipment run under raised floors. These cables are often very rigid and difficult to install. Adding new cables is also difficult. Again, the small size and resulting flexibility of fiber eliminate such problems. Indeed, in some cases, so few fibers are needed that the need for a raised floor can also be eliminated.

SAFETY

A fiber is a dielectric—it does not carry electricity. It presents no spark or fire hazard, so it cannot cause explosions or fires as a faulty copper cable can. Furthermore, it does not attract lightning. The fiber-optic cable can be run through hazardous areas, where electrical codes or common sense preclude the use of wires. It is possible, for example, to run a fiber directly through a fuel tank.

CONCLUSION

High bandwidth, low loss, and electromagnetic immunity are probably the three most outstanding features of fiber optics. These features dovetail nicely. They allow high-speed data transfer over long distances with little error. A fiber-optic link is capable of transmitting the entire text of a 30-volume encyclopedia over 100 miles in 1 s. The level of error is only one or two incorrect letters during the transmission.

Realize, however, that not all fibers have low loss and high bandwidth. Where loss and high speeds are not critical, as in automobiles, for example, less-expensive fibers work well. In an automobile, the central concern is protecting against noise from sources such as the ignition system. The other advantageous features of fiber optics make a well-suited transmission medium in many applications.

SUMMARY

- Fiber optics offers many advantages over copper cables.
- Optical fibers offer greater bandwidth.
- Optical fibers offer lower losses.
- High bandwidth and low losses mean greater repeater spacing in long-distance systems.
- Fibers offer excellent protection against EMI.
- Fibers are a secure transmission medium.
- Fibers are smaller and lighter than comparable copper cables.
- Since fibers do not carry electricity and cannot spark or cause fires, they are safe even in hostile environments.
- Fibers are used for different reasons, such as low loss, high bandwidth, security, and EMI immunity.

REVIEW QUESTIONS

1. List six advantages of fiber optics. Give an example of each.
2. Give three examples of rates and capacities in the North American Digital Telephone Hierarchy.
3. Which of the following is the most important reason for large bandwidth in optical fibers:
   A. High-speed, high-capacity transmission
   B. Secure transmission
   C. Fewer repeaters
   D. Immunity from electromagnetic interference
4. Does loss increase, decrease, or stay the same as the signal frequency increases in a copper cable? In an optical cable?

5. What is the name of the phenomenon that causes ghosts on your television, crackles on your radio, and other malfunctions in electronic equipment? Why does an optical fiber not contribute to this phenomenon?
6. For a 200-kHz signal traveling 50 m through a factory, which of the following factors are of great importance? Of little importance? Explain your reasons.
   A. Bandwidth
   B. Low loss
   C. EMI immunity
   D. Safety
   E. Error-free transmission
7. The original interest in optical fibers as a communication medium arose because
   A. They used less space than a coaxial cable.
   B. They were less expensive than copper.
   C. They could carry the high frequencies of laser light with little loss.
   D. Their electromagnetic immunity, security, and compatibility with digital techniques made them of use to the military.
8. (True/False) Information-carrying capacity depends on the size of the medium carrying the capacity.

Chapter 4

Light

Light is electromagnetic energy, as are radio waves, radar, television and radio signals, x-rays, and electronic digital pulses. *Electromagnetic energy* is radiant energy that travels through free space at about 300,000 km/s or 186,000 miles/s. An electromagnetic wave consists of oscillating electric and magnetic fields at right angles to each other and to the direction of propagation. Thus, an electromagnetic wave is usually depicted as a sine wave, as shown in Figure 4–1.

The main distinction between different waves lies in their frequency or wavelength. Frequency, of course, defines the number of sine-wave cycles per second and is expressed in hertz. *Wavelength* is the distance between the same points on two consecutive waves (or it is the distance a wave travels in a single cycle). Wavelength and frequency are related. Wavelength λ (lambda) equals the velocity of the wave (v) divided by its frequency (f):

$$\lambda = \frac{v}{f}$$

In free space or air, the velocity of an electromagnetic wave is the speed of light.

The equation clearly shows that the higher the frequency, the shorter the wavelength. For example, the 60-Hz power delivered to your house has a wavelength of 3100 miles. A 55.25-MHz signal, which carries the picture for channel 2 on television, has a wavelength of 17.8 ft. Deep red light has a frequency of

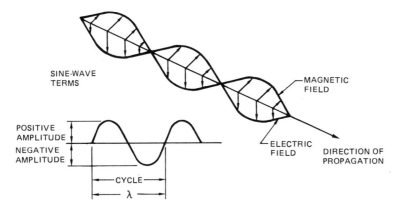

FIGURE 4–1 An electromagnetic wave

430,000 GHz (or 430 trillion hertz) and a wavelength of only 700 nm (nanometers or billionths of a meter).

In electronics, we customarily talk in terms of frequency. In fiber optics, however, light is described by wavelength. Remember, though, that frequency and wavelength are inversely related.

THE ELECTROMAGNETIC SPECTRUM

Electromagnetic energy exists in a continuous spectrum from subsonic energy through radio waves, microwaves, γ (gamma) rays, and beyond. (Electromagnetic energy exists in the subsonic range of the spectrum; however, sound itself is not electromagnetic energy but a vibration of air.) Figure 4–2 shows the spectrum. Notice that the radio frequencies most commonly used for communication are well below those of light.

$$\text{WAVELENGTH} = \frac{\text{SPEED OF LIGHT}}{\text{FREQUENCY}}$$

FIGURE 4–2 The electromagnetic spectrum (Illustration courtesy of AMP Incorporated)

Thus, light is electromagnetic energy with a *higher* frequency and *shorter* wavelength than radio waves. The figure also shows that the visible light we can see is only a small part of the light spectrum. Visible light has wavelengths from 380 nm for deep violet to 750 nm for deep red. Infrared light has longer wavelengths (lower frequencies) than visible light, whereas ultraviolet light has shorter ones. Most fiber-optic systems use infrared light between 800 and 1500 nm because glass fibers carry infrared light more efficiently than visible light.

The high frequencies of light have made it of such interest to communications engineers. As we saw in Chapter 2, a higher-frequency carrier means greater information-carrying capacity. Fiber optics is a method of using this information-carrying potential of light.

WAVES AND PARTICLES

So far, we have described light as an oscillating electromagnetic wave. It is spread out in space, without a definite, discrete location. Physicists once divided all matter into either waves or particles. We usually think of light as a wave and an electron as a particle. Modern physicists, however, have shown that this distinction does not exist. Both light and electrons exhibit wavelike and particlelike traits.

A particle of light is called a *photon,* which is a *quantum* or bundle of energy. A quantum exists in discrete units of energy—you cannot have half a quantum or 5.33 quanta. The amount of energy possessed by a photon depends on its frequency. The amount of energy increases as the frequency increases: Higher frequency means more energy. Wavelengths of violet light have more energy than those of red light because they have higher frequencies. The energy *E,* in watts, contained in a photon is

$$E = hf$$

where *f* is its frequency and *h* is Planck's constant, which is 6.63×10^{-34} J-s (joule-seconds). The equation shows clearly that differences in photon energy are strictly functions of frequency (or wavelength). Photon energy is proportional to frequency.

The photon is actually a strange particle, for it has zero rest mass. If it is not in motion, it does not exist! In this sense, it is not a particle as, say, marbles, stones, or ink drops are particles. It is a bundle of energy that acts like a particle.

Treating light as both a wave and a particle aids investigation of fiber optics. We switch easily between the two descriptions, depending on our needs. For example, many characteristics of optical fibers vary with wavelength, so the wave description is used. On the other hand, the emission of light by a source or its absorption by a detector is best treated by particle theory. The description of a detector speaks of light photons striking the detector and being absorbed. This absorption provides the energy required to set electrons flowing as current. A light-emitting diode (LED) operates because its electrons give up energy as photons. The exact energy of the photon determines the wavelength of the emitted light.

LIGHT RAYS AND GEOMETRIC OPTICS

The simplest way to view light in fiber optics is by ray theory. The light is treated as a simple ray, shown by a line. An arrow on the line shows the direction of propagation. The movement of light through the fiber-optic system can be analyzed with simple geometry. This approach not only simplifies analysis, but it also makes the operation of an optical fiber simple to understand.

REFLECTION AND REFRACTION

What is commonly called the *speed of light* is actually the velocity of electromagnetic energy in a vacuum such as space. Light travels at slower velocities in other materials such as glass. Light traveling from one material to another changes speed, which results in light changing its direction of travel. This deflection of light is called *refraction.* In addition, different wavelengths of light travel at different speeds in the *same* material. The variation of velocity with wavelength plays an important role in fiber optics.

Fishermen are quite familiar with refraction, since it distorts the apparent position of a fish lying stationary underwater (Figure 4–3). If one stands on a pier and looks directly down on the fish, the light is not refracted so that the fish is where it appears to be. If the fish is viewed by looking into the water at an angle, refraction of light occurs. What appears to be a straight line from the fish to the eye is actually a line with a bend where the light passes from water into air and is refracted. As a result, the fish is actually deeper in the water than it appears to be.

The prism in Figure 4–4 also demonstrates refraction. White light entering the prism contains all colors. The prism refracts the light, and it changes speed as it enters the prism. Because each wave changes speed differently, each is refracted differently. Red light deviates the least and travels the fastest. Violet light deviates the most and travels the slowest. The light emerges from the prism

FIGURE 4-3 Refraction

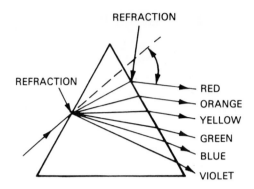

FIGURE 4-4 Refraction and a prism (Illustration courtesy of AMP Incorporated)

divided into the colors of the rainbow. Notice that refraction occurs at the entrance and at the exit of the prism.

The index of refraction, symbolized by *n*, is a dimensionless number expressing the ratio of the velocity of light *c* in free space to its velocity *v* in a specific material:

$$n = \frac{c}{v}$$

Table 4-1 lists some representative indices of refractions for selected materials as well as the approximate speed of light through the materials.

Of particular importance to fiber optics is that the index of glass can be changed by controlling its composition.

The amount that a ray of light is refracted depends on the refractive indices of the two materials. But before looking at the mechanics of refraction, we must first define some terms vital to our discussion. Figure 4-5 illustrates several important terms to understanding light and its refraction.

MATERIAL	INDEX (*n*)	LIGHT VELOCITY (km/s)
Vacuum	1.0	300,000
Air	1.0003 (1)	300,000
Water	1.33	225,000
Fused quartz	1.46	205,000
Glass	1.5	200,000
Diamond	2.0	150,000
Silicon	3.4	88,000
Gallium arsenide	3.6	83,000

TABLE 4-1 Indices of refraction for various materials

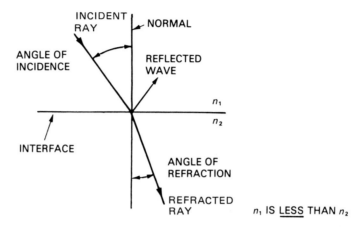

FIGURE 4-5 Angles of incidence and refraction (Illustration courtesy of AMP Incorporated)

- The *normal* is an imaginary line perpendicular to the interface of the two materials.
- The *angle of incidence* is the angle between the incident ray and the normal.
- The *angle of refraction* is the angle between the refracted ray and the normal.

Light passing from a lower refractive index to a higher one is bent *toward* the normal. But light going from a higher index to a lower one refracts *away* from the normal, as shown in Figure 4–6. As the angle of incidence increases, the angle of refraction approaches 90° to the normal. The angle of incidence that yields an angle of refraction of 90° is the *critical* angle. If the angle of incidence increases

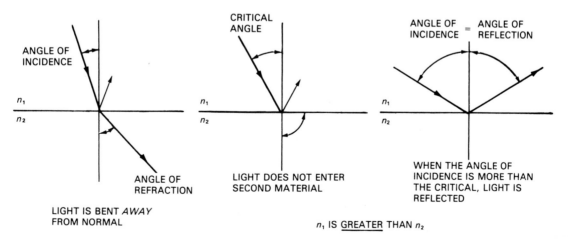

FIGURE 4-6 Reflection (Ilustration courtesy of AMP Incorporated)

past the critical angle, the light is totally reflected back into the first material so that it does not enter the second material. The angles of incidence and reflection are equal.

FRESNEL REFLECTIONS

Even when light passes from one index to another, a small portion is always reflected back into the first material. These reflections are *Fresnel reflections.* A greater difference in the indices of the materials results in a greater portion of the light reflecting. Fresnel reflection (ρ) at the boundary between air and another equals

$$\rho = \left(\frac{n-1}{n+1} \right)^2$$

In decibels, this loss of transmitted light is

$$dB = 10 \log_{10} (1 - \rho)$$

For light passing from air to glass (with $n = 1.5$ for glass), Fresnel reflection is about 0.17 dB. This figure will vary somewhat as the composition of the glass varies. Since such losses occur when light enters or exits an optical fiber, the loss in joining one fiber to another is 0.34 dB. A Fresnel reflection occurs when the light passes from the first fiber into the air gap separating the two fibers. A second Fresnel reflection occurs when the light passes from the air into the second fiber. Fresnel reflection is the same regardless of the order of materials through which the light passes; in other words, Fresnel reflection is the same as light passes from glass to air or from air to glass.

SNELL'S LAW

Snell's law states the relationship between the incident and refracted rays:

$$n_1 \sin \theta_1 = n_2 \sin \theta_2$$

The law shows that the angles depend on the refractive indices of the two materials. Knowing any three of the values, of course, allows us to calculate the fourth through simple rearrangement of the equation.

The critical angle of incidence θ_c, where $\theta_2 = 90°$, is

$$\theta_c = \arcsin \left(\frac{n_2}{n_1} \right)$$

At angles greater than θ_c, the light is reflected. Because reflected light means that n_1 and n_2 are equal (since they are in the same material), θ_1 and θ_2 are also equal. The angles of incidence and reflection are equal. These simple principles of refraction and reflection form the basis of light propagation through an optical fiber.

A PRACTICAL EXAMPLE

Figure 4–7 shows an example of reflection that has practical application in fiber optics. Assume we have two layers of glass, as shown in Figure 4–7A. The first layer, n_1, has a refractive index of 1.48; the second, 1.46. These values are typical for optical fibers. Using Snell's law, we can calculate the critical angle:

$$\theta_c = \arcsin\left(\frac{1.46}{1.48}\right)$$
$$= \arcsin(0.9864)$$
$$= 80.6°$$

Light striking the boundary between n_1 and n_2 at an angle greater than 80.6° from the normal reflects back into the first material. Again, the angle of incidence equals the angle of reflection.

Figure 4–7B carries the example a step further. Assume that a third layer of glass—labeled n_3—with the same refractive index as n_2 is placed on top of material n_1. Material n_1 is sandwiched between the n_2 and n_3 materials. We again have the same boundary condition as before. The reflected ray, however, now becomes the incident ray at the new boundary. The critical angle remains 80.6°. Thus all conditions are the same as in the first reflection. As a result, the light is again reflected back into the first material. The light reflected from n_3 becomes the incident ray again for n_2. The situation is the same. We have trapped the light between the two layers n_2 and n_3. As long as the angle of incidence is greater than 80.6°, the light is reflected back into the first material. Snell's law shows that, for this ideal example, this will always hold true. The light will continue traveling through the first material by total internal reflection.

(A)

(B)

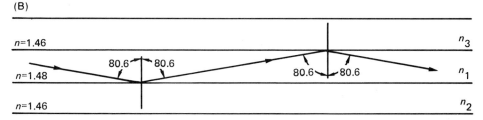

FIGURE 4–7 A practical example of reflection

The same principle explains the operation of an optical fiber. The main difference is that the fiber uses a circular configuration, such that n_2 surrounds n_1. The next chapter looks at light propagation in a fiber in detail.

SUMMARY

- Light is electromagnetic energy with a higher frequency and shorter wavelength than radio waves.
- Light has both wavelike and particlelike characteristics.
- When light meets a boundary separating materials of different refractive indices, it is either refracted or reflected.
- Fresnel reflections occur regardless of the angle of incidence.
- Snell's law describes the relationship between incident and reflected light.

REVIEW QUESTIONS

1. Describe the frequency and wavelength of light in relation to radio frequencies.
2. What is the name given to a particle of light?
3. Sketch a light ray being refracted. Label the incident ray, the refracted ray, and the normal.
4. Sketch a light ray being reflected. Label the incident ray, the reflected ray, and the normal.
5. If an incident ray passing from air into water has a 75° angle with respect to the normal, what is the angle of refracted ray?
6. What is the wavelength of a 300-MHz electromagnetic wave in free space? A 250-kHz wave? A 2-GHz wave?
7. Define Fresnel reflection.
8. Define critical angle.
9. Does light travel faster in air or in glass?

Part Two

FIBER-OPTIC COMPONENTS

Chapter 5

The Optical Fiber

The last chapter showed the characteristics of light propagation most important to your understanding to fiber optics. We saw that the reflection or refraction of light depends on the indices of refraction of the two media and on the angle at which light strikes the interface. The optical fiber works on these principles. Once light begins to reflect down the fiber, it will continue to do so under normal circumstances. The purpose of this chapter is to describe the propagation of light through the various types of optical fibers. The next chapter further develops the properties of fibers.

Keep in mind the distinction between the optical fiber and the fiber-optic cable. The *optical fiber* is the signal-carrying member, similar in function to the metallic conductor in a wire. But the fiber is usually cabled—that is, placed in a protective covering that keeps the fiber safe from environmental and mechanical damage. This chapter deals specifically with the optical fiber itself.

BASIC FIBER CONSTRUCTION

The optical fiber has two concentric layers called the *core* and the *cladding*. The inner core is the light-carrying part. The surrounding cladding provides the difference in refractive index that allows total internal reflection of light through the core. The index of the cladding is less than 1% lower than that of the core. Typical values, for example, are a core index of 1.47 and a cladding index of 1.46. Fiber manufacturers must carefully control this difference to obtain desired fiber characteristics.

Most fibers have an additional coating around the cladding. The coating, which is usually one or more layers of polymer, protects the core and cladding from shocks that might affect their optical or physical properties. The coating has no optical properties affecting the propagation of light within the fiber. This buffer coating, then, is a shock absorber.

Figure 5–1 shows the idea of light traveling through a fiber. Light injected into the fiber and striking the core-to-cladding interface at greater than the critical angle reflects back into the core. Since the angles of incidence and reflection are equal, the reflected light will again be reflected. The light will continue zigzagging down the length of the fiber.

FIGURE 5-1 Total internal reflection in an optical fiber (Illustration courtesy of AMP Incorporated)

Light, however, striking the interface at less than the critical angle passes into the cladding, where it is lost over distance. The cladding is usually inefficient as a light carrier, and light in the cladding becomes attenuated fairly rapidly.

Notice also, in Figure 5-1, that the light is also refracted as it passes from air into the fiber. Thereafter, its propagation is governed by the indices of the core and cladding and by Snell's law.

Such total internal reflection forms the basis of light propagation through a simple optical fiber. This analysis, however, considers only *meridional rays*— those that pass through the fiber axis each time they are reflected. Other rays, called *skew rays*, travel down the fiber without passing through the axis. The path of a skew ray is typically helical, wrapping around and around the central axis. Fortunately, skew rays are ignored in most fiber-optic analyses.

The specific characteristics of light propagation through a fiber depends on many factors, including

- The size of the fiber
- The composition of the fiber
- The light injected into the fiber

An understanding of the interplay between these properties will clarify many aspects of fiber optics.

Fibers themselves have exceedingly small diameters. Figure 5-2 shows cross sections of the core and cladding diameters of four commonly used fibers. The diameters of the core and cladding are as follows:

Core (μm)	Cladding (μm)
8	125
50	125
62.5	125
100	140

FIGURE 5-2 Typical core and cladding diameters

To realize how small these sizes are, note that human hair has a diameter of about 100 μm. Fiber sizes are usually expressed by first giving the core size, followed by the cladding size: thus, 50/125 means a core diameter of 50 μm and a cladding diameter of 125 μm; 100/140 means a 100-μm core and a 140-μm cladding. Through these small sizes are sent thousands of telephone conversations.

FIBER CLASSIFICATION

Optical fibers are classified in two ways. One way is by their material makeup:

- Glass fibers have a glass core and glass cladding. Since they are by far the most widely used, most discussion in this book centers on glass fibers. The glass used in fibers is ultrapure, ultratransparent silicon dioxide or fused quartz. If seawater were as clear as a fiber, you could see to the bottom of the deepest ocean trench, the 32,177-foot-deep Mariana Trench in the Pacific. Impurities are purposely added to the pure glass to achieve the desired index of refraction. Germanium or phosphorous, for example, increase the index. Boron or fluorine decrease the index. Other impurities not removed when the glass is purified also remain. These, too, affect fiber properties by increasing attenuation by scattering or absorbing light.
- Plastic-clad silica (PCS) fibers have a glass core and plastic cladding. Their performance, though not as good as all-glass fibers, is quite respectable.
- Plastic fibers have a plastic core and plastic cladding. Compared with other fibers, plastic fibers are limited in loss and bandwidth. Their very low cost and easy use, however, makes them attractive in applications where high bandwidth or low loss are not a concern. Their electromagnetic immunity and security allow plastic fibers to be beneficially used.

Plastic and PCS fibers do not have the buffer coating surrounding the cladding.

The second way to classify fibers is by the refractive index of the core and the modes that the fiber propagates. Figure 5-3, which depicts the differences in fibers classified this way, shows three important ideas about fibers.

First, it shows the difference between the input pulse injected into a fiber and the output pulse emerging from the fiber. The decrease in height of the pulse

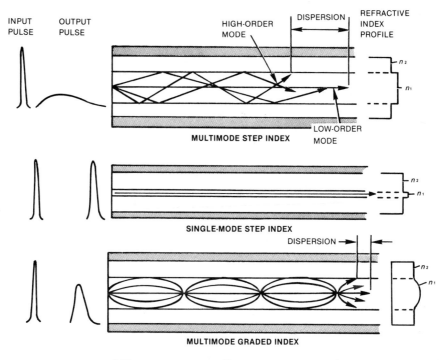

INPUT PULSE OUTPUT PULSE HIGH-ORDER MODE DISPERSION REFRACTIVE INDEX PROFILE

n_2

n_1

LOW-ORDER MODE

MULTIMODE STEP INDEX

n_2

n_1

SINGLE-MODE STEP INDEX

DISPERSION

n_2

n_1

MULTIMODE GRADED INDEX

FIGURE 5–3 Types of fiber propagation (Illustration courtesy of AMP Incorporated)

shows loss of signal power. The broadening in width limits the fiber's bandwidth or signal-carrying capacity. Second, it shows the path followed by light rays as they travel down the fiber. Third, it shows the relative index of refraction of the core and cladding for each type of fiber. The significance of these ideas will become apparent as we examine each type of fiber.

MODES

Mode is a mathematical and physical concept describing the propagation of electromagnetic waves through media. In its mathematical form, mode theory derives from Maxwell's equations. James Clerk Maxwell, a Scottish physicist in the last century, first gave mathematical expression to the relationship between electric and magnetic energy. He showed that they were both a single form of electromagnetic energy, not two different forms as was then believed. His equations also showed that the propagation of this energy followed strict rules. Maxwell's equations form the basis of electromagnetic theory.

A mode is an allowed solution to Maxwell's equations. For purposes of this book, however, a mode is simply a path that a light ray can follow in traveling down a fiber. The number of modes supported by a fiber ranges from 1 to over

100,000. Thus, a fiber provides a path of travels for one or thousands of light rays, depending on its size and properties.

In addition, each mode carries a characteristic amount of optical energy. Most fibers today support many modes. When light is first injected into the fiber, a mode may carry too much or too little energy, depending on the injected light. Some modes may carry no energy (meaning that no light travels along that potential path). Over distance, however, energy transfers between modes until all modes carry their characteristic energy. When a fiber reaches this point, it is said to have reached the steady state or equilibrium mode distribution (EMD). Achieving EMD in plastic fibers requires only a few meters or less. For high-performance glass fibers, EMD often requires tens of kilometers.

Because modes do not always carry the same amount of energy, fibers are spoken of as being overfilled or underfilled. In an *overfilled* fiber, all modes are excited. In addition, the source may launch some light into the cladding. Since, however, the cladding is not meant to carry light efficiently, these cladding modes will soon be lost. The fiber is overfilled: It carries energy in more modes than it will at EMD. The opposite is true of an underfilled fiber. Only some modes carry energy at first. Over distance, other modes will receive energy through mode coupling. As discussed in Chapter 11, such modal conditions can significantly affect real and apparent link performances. For now, it is important to understand the ideas of underfilled modes, overfilled modes, and EMD.

REFRACTIVE INDEX PROFILE

The *refractive index profile* describes the relation between the indices of the core and cladding. Two main relationships exist: step index and graded index. The step-index fiber has a core with a uniform index throughout. The profile shows a sharp step at the junction of the core and cladding. In contrast, the graded index has a nonuniform core. The index is highest at the center and gradually decreases until it matches that of the cladding. There is no sharp break between the core and the cladding.

By this classification, there are three types of fibers (whose names are often shortened by a prevalent characteristic):

1. Multimode step-index fiber (commonly called step-index fiber)
2. Multimode graded-index fiber (graded-index fiber)
3. Single-mode step-index fiber (single-mode fiber)

The characteristics of each type have important bearing on its suitability for particular applications. As we progress through this chapter and the next, the importance of each type will become apparent.

STEP-INDEX FIBER

The multimode step-index fiber is the simplest type. It has a core diameter from 100 to 970 μm, and it includes glass, PCS, and plastic constructions. As such,

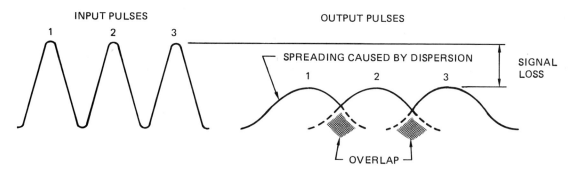

FIGURE 5-4   Pulse spreading

the step-index fiber is the most wide ranging, although not the most efficient in having high bandwidth and low losses.

Since light reflects at different angles for different paths (or modes), the path lengths of different modes are different. Thus, different rays take a shorter or longer time to travel the length of the fiber. The ray that goes straight down the center of the core without reflecting arrives at the other end first. Other rays arrive later. Thus, light entering the fiber at the same time exits the other end at different times. The light has spread out in time.

This spreading of light is called *modal dispersion.* Dispersion describes the spreading of light by various mechanisms. Modal dispersion is that type of dispersion that results from the varying modal path lengths in the fiber. Typical modal dispersion figures for the step-index fiber are 15 to 30 ns/km. This means that for light entering a fiber at the same time, the ray following the longest path will arrive at the other end of a 1-km-long fiber 15 to 30 ns after the ray following the shortest path.

Fifteen to 30 billionths of a second may not seem like much, but dispersion is the main limiting factor on a fiber's bandwidth. Pulse spreading results in a pulse overlapping adjacent pulses, as shown in Figure 5-4. Eventually, the pulses will merge so that one pulse cannot be distinguished from another. The information contained in the pulse is lost. Reducing dispersion increases fiber bandwidth.

GRADED-INDEX FIBER

One way to reduce modal dispersion is to use graded-index fibers. Here the core has numerous concentric layers of glass, somewhat like the annular rings of a tree. Each successive layer outward from the central axis of the core has a lower index of refraction. Figure 5-5 shows the core's structure.

Light, remember, travels faster in a lower index of refraction. So the further the light is from the center axis, the greater is its speed. Each layer of the core refracts the light. Instead of being sharply reflected as it is in a step-index fiber, the light is now bent or continually refracted in an almost sinusoidal pattern. Those rays that follow the longest path by traveling near the outside of the core

FIGURE 5–5 Concentric layers of lower refractive index in a graded-index core (Courtesy of AT & T Bell Laboratories)

have a faster average velocity. The light traveling near the center of the core has the slowest average velocity. As a result, all rays tend to reach the end of the fiber at the same time. The graded index reduces modal dispersion to 1 ns/km or less.

Popular graded-index fibers have core diameters of 50, 62.5, or 85 μm and a cladding diameter of 125 μm. The fiber is popular in applications requiring a wide bandwidth, especially telecommunications.

SINGLE-MODE FIBER

Another way to reduce modal dispersion is to reduce the core's diameter until the fiber only propagates one mode efficiently. The single-mode fiber has an exceedingly small core diameter of only 5 to 10 μm. Standard cladding diameter is 125 μm. This cladding diameter was chosen for three reasons:

1. The cladding must be about 10 times thicker than the core in a single-mode fiber. For a fiber with an 8-μm core, the cladding should be at least 80 μm.
2. It is the same size as graded-index fibers, which promotes size standardization.
3. It promotes easy handling because it makes the fiber less fragile and because the diameter is reasonably large so that it can be handled by technicians.

Since this fiber carries only one mode, modal dispersion does not exist.

Single-mode fibers easily have a potential bandwidth of 50 to 100 GHz-km. Present fibers have a bandwidth of several gigahertz and allow transmission of tens of kilometers. As of early 1985, the largest commercially available fiber-optic systems were digital telephone transmission systems operating around 417 Mbps. These systems carried 6048 simultaneous telephone calls over a single-mode fiber a distance of 35 km without a repeater. By the end of 1985, capacities had grown fourfold to 1700 Mbps and 24,192 voice channels.

An important aspect of this growth is that the increase results from changing the electronics, not the single-mode fibers, at either end of the system. The capacity of a single-mode system is limited by the capabilities of the electronics, not of the fiber. One advantage of single-mode fibers is that once they are installed, the system's capacity can be increased as newer, higher-capacity transmission electronics becomes available. This capability saves the high cost of installing a new transmission medium to obtain increased performance and allows cost-effective increases from 417 to 1700 Mbps.

The point at which a single-mode fiber propagates only one mode depends on the wavelength of light carried. A wavelength of 820 nm results in multimode operation. As the wavelength is increased, the fiber carries fewer and fewer modes until only one remains. Single-mode operation begins when the wavelength approaches the core diameter. At 1300 nm, for example, the fiber permits only one mode. It becomes a single-mode fiber.

Different fiber designs have a specific wavelength, called the *cutoff* wavelength, above which it carries only one mode. A fiber designed for single-mode operation at 1300 nm has a cutoff wavelength of around 1200 nm.

The operation of a single-mode fiber is slightly more complex than simply a ray traveling down the core. Geometric optics using light rays is not as appropriate for these fibers because it obscures how optical energy is distributed within the fiber. Some of the optical energy of the mode travels in the cladding, as shown in Figure 5-6. Therefore, the diameter of the light appearing at the end of

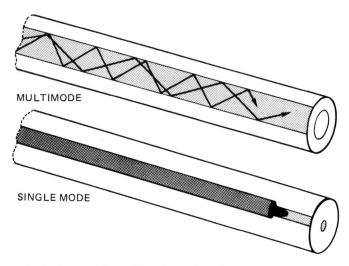

MULTIMODE

SINGLE MODE

FIGURE 5-6 Optical power in multimode and single-mode fibers (Courtesy of Corning Glass Works)

the fiber is larger than the core diameter. *Mode field diameter* is the term used to define this diameter of optical energy. Although optical energy is confined to the core in a multimode fiber, it is not so confined in a single-mode fiber. It is usually more important to know the mode field diameter than the core diameter.

The difference in propagation of light in a single-mode fiber points to another difference between single-mode and multimode fibers. Because optical energy in a single-mode fiber travels in the cladding as well as in the core, the cladding must be a more efficient carrier of energy. In a multimode fiber, the light transmission characteristics of the cladding are basically unimportant. Indeed, because cladding modes are not desirable, a cladding with inefficient transmission characteristics can be tolerated. This situation does not hold for a single-mode fiber.

ADVANCED SINGLE-MODE DESIGNS

Not all single-mode fibers use a step profile. Some use more complex designs in order to optimize the fiber to operate at a certain wavelength. For example, a step-index fiber has zero material dispersion at 1300 nm. Zero dispersion—discussed further in the next chapter—is important because it is at this wavelength that a fiber has its greatest information-carrying capacity. Dispersion is about five times higher at 1550 nm. Attenuation levels, however, are significantly lower at 1550 nm:

- 0.35 to 0.50 dB/km at 1300 nm
- 0.20 to 0.30 dB/km at 1550 nm

The differences in attenuation and dispersion characteristics at the two wavelengths represent a tradeoff to system designers. They can operate at higher speeds and shorter distances at 1300 nm, or they can operate at lower speeds and longer distances at 1550 nm.

The advanced core designs of newer single-mode fibers attempt to make lower loss and low dispersion coincide at the same wavelength, so the system can operate at very high speeds over very long distances. Dispersion-shifted fibers have a structure that changes the zero-dispersion wavelength, typically from 1300 to 1550 nm. Dispersion-flattened fibers have a structure that lowers dispersion over a wide range of wavelengths. Figure 5–7 shows the profiles of some typical advanced single-mode fiber profiles.

HOW MANY MODES ARE THERE?

We have seen that the number of modes supported by a fiber in part determines its information-carrying capacity. Modal dispersion, which causes pulse spreading and overlapping, limits the data rate that a fiber can support. We have also seen that dispersion depends on wavelength and core diameter.

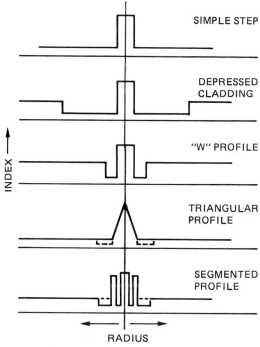

FIGURE 5-7 Refractive index profiles of advanced single-mode fibers (Courtesy of Corning Glass Works)

The V number, or normalized frequency, is a fiber parameter that takes into account the core diameter, wavelength propagated, and fiber NA (a fiber property described in the next chapter):

$$V = \frac{2\pi d}{\lambda} \ (NA)$$

From the V number, the number of modes in a fiber can be calculated.

For a simple step-index fiber, the number of modes can be approximated by

$$N = \frac{V^2}{2}$$

For a graded-index fiber, the number of modes can be approximated by

$$N = \frac{V^2}{4}$$

The equations demonstrate that the number of modes is determined by core diameter, fiber NA, and the wavelength propagated. The number of modes in a graded-index fiber is about half that of a step-index fiber having the same diameter and NA. A fiber with a 50-μm core supports over 1000 modes.

When the *V* number of a step-index fiber becomes 2.405, the fiber supports a single mode. The *V* number can be decreased by decreasing the core diameter, by increasing the operating wavelength, or by decreasing the NA. Thus, single-mode operation in a fiber can be obtained by suitably adjusting these characteristics.

FIBER COMPARISONS

Table 5–1 shows typical characteristics of various fibers. The meaning of terms such as NA are explained in the next chapter. For now, though, you can see that the performance and physical properties cover a broad range. We use the term "performance" broadly: Better performance means higher bandwidth, higher information-carrying capacity, and lower losses. Other measures of performance, such as safety or low cost, would favor other types of fibers. The table also suggests some generalizations about loss and bandwidth:

- Fiber performance from lowest to highest is as follows:
  —Plastic
  —PCS
  —Step-index glass
  —Graded index
  —Single mode
- A smaller core usually means better performance.
- Glass fibers perform better than plastic ones.

Remember, though, that such generalizations suffer the fault of all generalizations: They do not tell the whole story. A fiber-optic cable must be matched to the requirements of the application. A system needing to transmit only a few thousand bits a second over a couple of meters can use plastic fibers to best

TYPE	CORE DIA (μm)	CLADDING DIA (μm)	BUFFER DIA (μm)	NA	BANDWIDTH (MHz-km)	ATTENUATION (dB/km)
Single mode	8 5	125 125	250 250		6 ps/km* 4 ps/km*	.5 @ 1300 nm .4 @ 1300 nm
Graded index	50 63 85 100	125 125 125 140	250 130 250 250	.20 .29 .26 .30	400 250 200 20	4 @ 850 nm 7 @ 850 nm 6 @ 850 nm 5 @ 850 nm
Step index	200 300	380 440	600 650	.27 .27	25 20	6 @ 850 nm 6 @ 850 nm
PCS	200 400 600	350 550 900	-- -- --	.30 .30 .40	20 15 20	10 @ 790 nm 10 @ 790 nm 6 @ 790 nm
Plastic	-- --	750 1000	-- --	.50 .50	20 20	400 @ 650 nm 400 @ 650 nm

* DISPERSION PER NANOMETER OF SOURCE WIDTH

TABLE 5–1 Examples of fiber specifications (Illustration courtesy of AMP Incorporated) advantage. Not only does plastic fiber cost less, but compatible components such as sources, detectors, and connectors also cost less. Using a single-mode fiber in such applications would be like using a Ferrari to go to the corner store. It is common to stress the low losses and high bandwidths of fibers, but not all applications need such performance. All the fiber types have their uses.

SUMMARY

- The three types of fibers are step index, graded index, and single mode.
- Dispersion is one factor limiting fiber performance. One motive of fiber design is to reduce dispersion by grading the index or by using a single-mode design.
- Core diameter gives a rough estimate of fiber performance: the smaller the core, the higher the bandwidth and the lower the loss.
- Fibers come in a variety of performances to suit different application needs.
- Mode field diameter is the diameter of optical energy carried in a single-mode fiber.
- Single-mode fibers use both step-index and more complex profiles.

REVIEW QUESTIONS

1. What are the two main parts of an optical fiber?
2. What are the three types of fibers according to material composition?
3. Light travels down a step-index fiber by what principle?
4. What usually happens to the performance of a fiber if the core diameter is reduced? If it is enlarged?
5. What is the path called that is followed by optical energy down a fiber?
6. What is the name of a fiber that allows only one path for the light down the fiber?
7. What is the name of a fiber with a core whose refractive index varies?
8. Is the refractive index of the core higher or lower than that of the cladding?
9. (True/False) Modal dispersion is constant at all wavelengths.
10. (True/False) One advantage of the graded-index fiber is that the wavelengths of low loss and minimum dispersion always coincide.

Chapter 6

Fiber Characteristics

This chapter examines the characteristics of optical fibers most important to users and designers. The chapter expands on the characteristics discussed in previous chapters and introduces new ones.

DISPERSION

As discussed in the last chapter, *dispersion* is the spreading of a light pulse as it travels down the length of an optical fiber. Dispersion limits the bandwidth or information-carrying capacity of a fiber. The bit rate must be low enough to ensure that pulses do not overlap. A lower bit rate means that the pulses are farther apart and, therefore, that greater dispersion can be tolerated. There are three main types of dispersion:

1. Modal dispersion
2. Material dispersion
3. Waveguide dispersion

MODAL DISPERSION

As shown in the last chapter, modal dispersion occurs only in multimode fibers. It arises because rays follow different paths through the fiber and consequently arrive at the other end of the fiber at different times. Modal dispersion can be reduced in three ways:

1. Use a smaller core diameter, which allows fewer modes. A 100-μm-diameter core allows fewer paths than a 200-μm-diameter core.
2. Use a graded-index fiber so that the light rays that follow longer paths also travel at a faster average velocity and thereby arrive at the other end of the fiber at nearly the same time as rays that follow shorter paths.
3. Use a single-mode fiber, which permits no modal dispersion.

MATERIAL DISPERSION

Different wavelengths (colors) also travel at different velocities through a fiber, even in the same mode. Earlier, we saw that the index of refraction is equal to

$$n = \frac{c}{v}$$

where c is the speed of light in a vacuum and v is the speed of the same wavelength in the material.

Each wavelength, though, travels at a different speed through a material, so the value of v in the equation changes for each wavelength. Thus, index of refraction changes according to the wavelength. Dispersion from this phenomenon is called *material dispersion* since it arises from material properties of the fiber. The amount of dispersion depends on two factors:

1. The range of light wavelengths injected into the fiber. A source does not normally emit a single wavelength; it emits several. This range of wavelengths, expressed in nanometers, is the spectral width of the source. An LED has a much wider spectral width than a laser—about 35 nm for an LED and 2 to 3 nm for a laser.

2. The center operating wavelength of the source. Around 850 nm, longer ("reddish") wavelengths travel faster than shorter ("bluish") ones. An 860-nm wave travels through glass faster than an 840-nm wave. At 1550 nm, however, the situation is reversed: The shorter wavelengths travel faster than the longer ones, a 1560-nm wave travels slower than a 1540-nm wave. At some point, the crossover must occur where the bluish and reddish wavelengths travel at the same speed. This crossover occurs around 1300 nm, the zero-dispersion wavelength. Figure 6–1 shows this idea. The length of the arrows suggest the speed of the wavelength; hence, a longer arrow is traveling faster and arriving earlier.

Figure 6–2 is a graph showing dispersion for a typical single-mode fiber. Dispersion is zero at 1300 nm. At wavelengths below 1300 nm, dispersion is negative, so wavelengths trail or arrive later. Above 1300 nm, wavelengths lead or arrive faster.

FIGURE 6-1 Material dispersion and wavelength

Material Dispersion

FIGURE 6-2 Material dispersion and the zero-dispersion wavelength (Courtesy of Corning Glass Works)

Material dispersion is of greater concern in single-mode systems. In a multimode system, modal dispersion is usually significant enough that modal dispersion is not a factor. In many cases, designers do not even concern themselves with modal dispersion. Speeds are too low, or distances are too short.

The 820-to-850-nm region of the spectrum is used as a transmission wavelength for many fiber-optic systems. In this region, material dispersion can be approximated as being roughly equal to 0.1 ns/nm of spectral width.

WAVEGUIDE DISPERSION

Waveguide dispersion, most significant in a single-mode fiber, occurs because optical energy travels in both the core and cladding, which have slightly different refractive indices. The energy travels at slightly different velocities in the core and cladding because of the slightly different refractive indices of the materials. Altering the internal structure of the fiber allows waveguide dispersion to be substantially changed, thus changing the specified overall dispersion of the fiber, which is one goal of the advanced single-mode designs discussed in the last chapter.

BANDWIDTH AND DISPERSION

Many fiber and cable manufacturers do not specify dispersion for their multimode offerings. Instead, they specify a figure of merit called the bandwidth-length product or simply bandwidth, given in megahertz-kilometers. A bandwidth of 400 MHz-km means that a 400-MHz signal can be transmitted for 1 km. It also means that the product of the frequency and the length must be 400 or less (BW × L = 400). In other words, you can send a lower frequency a longer distance:

- 200 MHz for 2 km
- 100 MHz for 4 km
- 50 MHz for 8 km

Conversely, a higher frequency can be sent a shorter distance:

- 600 MHz for 0.66 km
- 800 MHz for 0.50 km
- 1000 MHz for 0.25 km

Single-mode fibers, on the other hand, are specified by dispersion. This dispersion is expressed in picoseconds per kilometer per nanometer of source spectral width (ps/km/nm). In other words, for any given single-mode fiber, dispersion is most affected by the source's spectral width: the wider the source width (the more wavelengths injected into the fiber), the greater the dispersion. Although the conversion of single-mode dispersion to bandwidth is complex, a rough approximation can be gained from the following equation:

$$BW = \frac{0.187}{(Disp)(SW)(L)}$$

where

Disp = dispersion at the operating wavelength, in seconds per nanometer per kilometer
SW = the spectral width (rms) of the source, in nanometers
L = fiber length, in kilometers

Here is an example using the following:

Dispersion = 3.5 ps/nm/km
Spectral width = 2 nm
Length = 25 km

Plugging these values into the equation yields

$$BW = \frac{0.187}{(3.5 \times 10^{-12}\,s/nm/km)(2\,nm)(25\,km)}$$
$$= 1068\ \text{MHz} \quad \text{or, roughly,} \quad 1\ \text{GHz}$$

Doubling the source width to 4 nm significantly reduces the bandwidth to around 535 MHz. So the spectral width of the source has a significant effect on the performance of a single-mode fiber. Reducing the dispersion figure or the source's spectral width increases the bandwidth.

ATTENUATION

Attenuation is the loss of optical power as light travels through the fiber. Measured in decibels per kilometer, it ranges from over 300 dB/km for plastic fibers to around 0.21 dB/km for single-mode fibers.

Attenuation varies with the wavelength of light. Windows are low-loss regions, where fibers carry light with little attenuation. The first generation of optical fibers operated in the first window, around 820 to 850 nm. The second window is the zero-dispersion region of 1300 nm, and the third window is the 1550-nm region. A typical 50/125 graded-index fiber offers attenuation of 4 dB/km at 850 nm and 2.5 dB/km at 1300 nm, which is about a 30% increase in transmission efficiency.

High-loss regions, where attenuation is very high, occur at 730, 950, 1250, and 1380 nm. One wishes to avoid operating in these regions. Evaluating loss in a fiber must be done with respect to the transmitted wavelength. Figure 6–3 shows a typical attenuation curve for a low-loss multimode fiber. Figure 6–4 does the same for a single-mode fiber; notice the high loss in the mode-transition region, where the fiber shifts from multimode to single-mode operation.

Making the best use of the low-loss properties of the fiber requires that the source emit light in the low-loss regions of the fiber.

Plastic fibers are best operated in the visible-light area around 650 nm.

One important feature of attenuation in an optical fiber is that it is constant at all modulation frequencies within the bandwidth. In copper cables, attenuation

FIBER SPECTRAL ATTENUATION

FIGURE 6–3 Attenuation versus wavelength for a multimode fiber (Courtesy of Corning Glass Works)

FIGURE 6-4 Attenuation versus wavelength for a single-mode fiber (Courtesy of Corning Glass Works)

increases with the frequency of the signal: the higher the frequency, the greater the attenuation. A 25-MHz signal will be attenuated in a copper cable more than will a 10-MHz signal. As a result, signal frequency limits the distance a signal can be sent before a repeater is needed to regenerate the signal. In an optical fiber, both signals will be attenuated the same.

Attenuation in a fiber has two main causes:

1. Scattering
2. Absorption

SCATTERING

Scattering is the loss of optical energy due to imperfections in the fiber and from the basic structure of the fiber. Scattering does just what the term implies: It scatters the light in all directions (Figure 6-5). The light is no longer directional.

Rayleigh scattering is the same phenomenon that causes a red sky at sunset. The shorter blue wavelengths are scattered and absorbed while the longer red wavelengths suffer less scattering and reach our eyes, so we see a red sunset.

Rayleigh scattering comes from density and compositional variations in a fiber that are natural by-products of manufacturing. Ideally, pure glass has a

FIGURE 6-5 Scattering

perfect molecular structure and, therefore, uniform density throughout. In real glass, the density of the glass is not perfectly uniform. The result is scattering.

Since scattering is inversely proportional to the fourth power of the wavelength ($1/\lambda^4$), it decreases rapidly at longer wavelengths. Scattering represents the theoretical lower limits of attenuation, which are as follows:

- 2.5 dB at 820 nm
- 0.24 dB at 1300 nm
- 0.012 dB at 1550 nm

ABSORPTION

Absorption is the process by which impurities in the fiber absorb optical energy and dissipate it as a small amount of heat. The light becomes "dimmer." The high-loss regions of a fiber result from water bands, where hydroxyl molecules significantly absorb light. Other impurities causing absorption include ions of iron, copper, cobalt, vanadium, and chromium. To maintain low losses, manufacturers must hold these ions to less than one part per billion. Fortunately, modern manufacturing techniques, including making fibers in very clean environments, permits control of impurities to the point that absorption is not nearly as significant as it was just a few years ago.

MICROBEND LOSS

Microbend loss is that loss resulting from microbends, which are small variations or "bumps" in the core-to-cladding interface. As shown in Figure 6–6, microbends can cause high-order modes to reflect at angles that will not allow further reflection. The light is lost.

Microbends can occur during the manufacture of the fiber, or they can be caused by the cable. Manufacturing and cabling techniques have advanced to minimize microbends and their effects.

NUMERICAL APERTURE

Numerical aperture (NA) is the "light-gathering ability" of a fiber. Only light injected into the fiber at angles greater than the critical angle will be propagated. The *material* NA relates to the refractive indices of the core and cladding:

$$NA = \sqrt{n_1^2 - n_2^2}$$

Notice that NA is a unitless dimension.

We can also define the angles at which rays will be propagated by the fiber. These angles form a cone, called the *acceptance cone*, that gives the maximum angle of light acceptance. The acceptance cone is related to the NA:

$$\theta = \arcsin (NA)$$
$$NA = \sin \theta$$

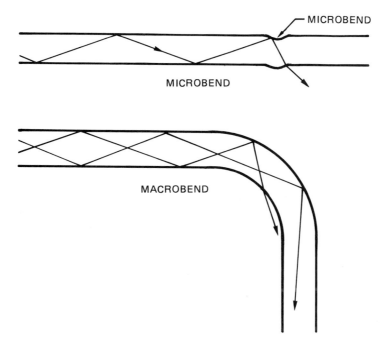

FIGURE 6-6   Loss and bends

where θ (theta) is the half-angle of acceptance. See Figure 6-7.

The NA of a fiber is important because it gives an indication of how the fiber accepts and propagates light. A fiber with a large NA accepts light well; a fiber with a low NA requires highly directional light.

In general, fibers with a high bandwidth have a lower NA. They thus allow fewer modes. Fewer modes mean less dispersion and, hence, greater bandwidth. NAs range from about 0.50 for plastic fibers to 0.21 for graded-index fibers. A large NA promotes more modal dispersion, since more paths for the rays are provided.

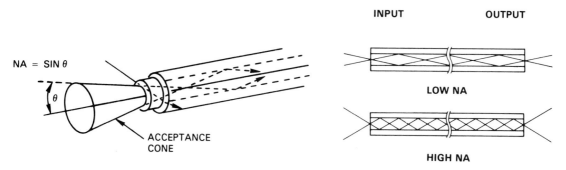

FIGURE 6-7   Numerical aperture (Illustration courtesy of AMP Incorporated)

FIGURE 6-8   Active area carrying optical power in a 62.5-μm core under fully filled and EMD conditions (Illustration courtesy of AMP Incorporated)

Manufacturers do not normally specify NA for their single-mode fibers (the NA is only about 0.11), because NA is not a critical parameter for the system designer or user. Light in a single-mode fiber is not reflected or refracted, so it does not exit the fiber at angles. Similarly, the fiber does not accept light rays at angles within the NA and propagate them by total internal reflection. As a result, NA, although it can be defined for a single-mode, is essentially meaningless as a practical characteristic. Later chapters will show how NA in a multimode fiber is important to system performance and to calculating anticipated performance.

The NA of a fiber changes over distance. High-order modes—those that travel near the critical angle—are often lost. As a graded-index fiber reaches EMD, for example, its NA can be reduced by as much as 50%. This reduction means that the light exiting the fiber does so at angles much less than defined by the acceptance cone. In addition, the spot diameter of light emerging from the fiber can also be reduced. Figure 6–8 shows the diameter of light in the core of a 62.5-μm core under fully filled and EMD conditions. When the fiber is fully filled, the light fills the core. When the fiber reaches EMD, the light diameter is only 50 μm. The NA of the light is similarly reduced.

Sources and detectors also have an NA. The NA of a source defines the angles of the exiting light. The NA of the detector defines the angles of light that will operate the detector. Especially for sources, it is important to match the NA of the source to the NA of the fiber so that all the light emitted by the source is coupled into the fiber and propagated. Mismatches in NA are sources of loss when light is coupled from a lower NA to a higher one.

FIBER STRENGTH

One usually thinks of glass as brittle. Certainly, a pane of glass is not easily bent, let alone rolled up. Yet, a fiber can be looped into tight circles without breaking. Furthermore, fiber can be tied into loose knots. (Pulling the knot tight will break the fiber.)

Tensile strength is the ability of a fiber or wire to be stretched or pulled without breaking. The tensile strength of a fiber exceeds that of a steel filament the same size. Furthermore, a copper wire must have twice the diameter to have the same tensile strength as a fiber.

The main causes of weakness in a fiber are microscopic cracks on the surface and flaws within the fiber. Surface cracks are probably most significant. Surface defects can grow with pulling from the tensile load applied as the fiber is installed and from the tensile load on the fiber during its installed lifetime. Temperature changes, mechanical and chemical damage, and normal aging also promote defects.

Defects can grow, eventually causing the fiber to break. If you have ever cut glass, you understand this phenomenon. To cut glass, you make a shallow scribe across the glass. Given a sharp snap, the glass splits along the scribe. The same effect holds true for optical fibers. A flaw acts like the line scribed in the pane of glass. As the fiber is pulled, the flaw grows into the fiber until the fiber breaks.

BEND RADIUS

Even though fibers can be wrapped in circles, they have a minimum bend radius. A sharp bend will snap the glass. Bends have two other effects:

1. They increase attenuation slightly. This effect should be intuitively clear. The bends change the angles of incidence and reflection enough that some high-order modes are lost (similar to microbends).
2. Bends decrease the tensile strength of the fiber. If pull is exerted across a bend, the fiber will fail at a lower tensile strength than if no bend were present.

As a rule of thumb, the minimum bend radius is five times the cable diameter for an *unstressed* cable and 10 times the diameter for a *stressed* cable.

NUCLEAR HARDNESS

Nuclear hardness refers to the ability of equipment to withstand nuclear effects. The effect of nuclear radiation on a conductor is of great interest to the U.S. military (in particular, to protect and maintain its command, control, and communication (C^3) system), to the nuclear power industry, and to other areas of high-radiation risk. Fibers are nonconductive and do not build up static charges when exposed to radiation. Fibers will not short out if their jackets are melted by the high heat of a nuclear emergency.

Fibers do suffer an increase in attenuation when subjected to high levels of continuous nuclear radiation. Nuclear radiation increases the light absorption by the fiber's impurities. The increase in attenuation depends on the amount of the dosage and the intensity of the dosage. A high-intensity burst of 3700 rad in 3 ns increases attenuation to peaks of thousands of decibels per kilometer. This excess attenuation decreases to under 10 dB/km in less than 10 s and to fewer than

5 dB/km in less than 100 s. Thus, glass fibers are able to transmit information after exposure to high-radiation weapons bursts within minutes after the radiation has cleared.

An electromagnetic pulse (EMP) is another concern associated with nuclear weapons, although its effects are more closely related to severe EMI than to nuclear radiation. An EMP results from a nuclear explosion. Two or three nuclear devices detonated over the United States at an altitude of several hundred kilometers could damage or destroy every piece of unprotected electronic equipment in the country.

Gamma rays, produced in the first few nanoseconds after detonation, travel until they collide with electrons of air molecules in the upper atmosphere. These electrons, scattered and accelerated by the collisions, are deflected by the earth's magnetic field to produce a transverse electric current. This current sets up EMPs that radiate downward. Any metal conductors will pick up these pulses and conduct them. A 1-megaton warhead can produce peak fields of 50 kV/m with an instantaneous peak power of around 6 MW/m². Such levels are far beyond the ability of equipment to withstand. In short, the entire power and communications grid of the country could be wiped out by an electromagnetic pulse.

SUMMARY

- Dispersion is a general term for phenomena that cause light to spread as it travels through a fiber.
- The three types of dispersion are modal, material, and waveguide.
- Dispersion limits bandwidth.
- Dispersion in multimode fibers includes modal and material dispersion.
- Dispersion in single-mode fibers includes material and waveguide dispersion. Material dispersion is the most important.
- Attenuation is the loss of signal power.
- Attenuation varies with the frequency of light.
- Attenuation does not vary with the signal rate in an optical fiber.
- Numerical aperture is the light-gathering ability of a fiber. It defines the angles at which a fiber will accept and propagate light.
- Fibers have a higher tensile strength than do wires of comparable size.

REVIEW QUESTIONS

1. Name three types of dispersion.
2. Which type of dispersion does not exist in a single-mode fiber?
3. Is the information-carrying capacity of a multimode fiber characterized by dispersion or by bandwidth?
4. Is the information-carrying capacity of a single-mode fiber characterized by dispersion or by bandwidth?

5. If a multimode fiber has a bandwidth of 250 MHz-km, how far can a 750-MHz signal be transmitted?
6. Name the two main mechanisms of attenuation in an optical fiber.
7. Is attenuation lower in a fiber at 850 nm, 1300 nm, or 1550 nm? Why?
8. For any particular single-mode fiber, what specific characteristic limits its data rate?
9. Does an optical fiber recover from a short, intense exposure from nuclear radiation in seconds, minutes, hours, or days?
10. (True/False) The material NA of a fiber always indicates the modal conditions of light traveling within the fiber.

Chapter 7

Fiber-Optic Cables

Most often, an optical fiber must be packaged before it is used. Packaging involves cabling the fiber. *Cabling* is an outer protective structure surrounding one or more fibers. It is analogous to the insulation or other materials surrounding a copper wire. Cabling protects copper and fibers environmentally and mechanically from being damaged or degraded in performance. (The additional protection against electric shock, shorts, and the possibility of fire important to copper is not an issue with dielectric fibers.) The purpose of this chapter is to describe some typical cable structures.

Like their copper counterparts, fiber-optic cables come in a great variety of configurations (Figure 7–1). Important considerations in any cable are tensile strength, ruggedness, durability, flexibility, environmental resistance, temperature extremes, and even appearance.

FIGURE 7–1   Fiber-optic cables (Photo courtesy of ITT Electro-Optical Products)

Evaluation of these considerations depends on the application. An outside-plant telephone cable must endure extreme hardship. It must withstand extremes of heat and cold, ice deposits that cause it to sag on a pole, high winds that buffet it, and rodents that chew on it underground. It obviously must be more rugged than a cable connecting equipment within the controlled environment of a telephone switching office. Similarly, a cable running under an office carpet, where people will walk on it and chairs will roll over it, has different requirements than a cable running within the walls of the office.

MAIN PARTS OF A FIBER-OPTIC CABLE

Figure 7–2 shows the main parts of a simple single-fiber cable. Even though cables come in many varieties, most have the following parts in common:

- Optical fiber
- Buffer
- Strength member
- Jacket

Since we have already looked extensively at optical fibers, we discuss only the buffer, strength member, and jacket here.

BLACK
POLYURETHANE
OUTER JACKET

STRENGTH
MEMBERS

BUFFER JACKET

SILICONE COATING

CLADDING (SILICA)

CORE (SILICA)

OPTICAL FIBER

FIGURE 7–2 Parts of a fiber-optic cable (Courtesy of Hewlett-Packard)

BUFFERS

The simplest buffer is the plastic coating applied to the cladding and discussed in Chapter 5. This buffer, which is part of the fiber, is applied by the fiber manufacturer. An additional buffer is added by the cable manufacturer. (Most cable manufacturers and vendors do not make their own fibers.)

The cable buffer is one of two types: loose buffer or tight buffer. Figure 7–3 shows the two constructions and summarizes the tradeoffs involved with each.

The *loose buffer* uses a hard plastic tube having an inside diameter several times that of the fiber. One or more fibers lie within the buffer tube. The tube isolates the fiber from the rest of the cable and the mechanical forces acting on it. The buffer becomes the load-bearing member. As the cable expands and shrinks with changes in temperature, it does not affect the fiber as much. A fiber has a lower temperature coefficient than most cable elements, meaning that it expands and contracts less. Typically, some excess fiber is in the tube; in other words, the fiber in the tube is slightly longer than the tube itself. Thus the cable can easily expand and contract without stressing the fiber.

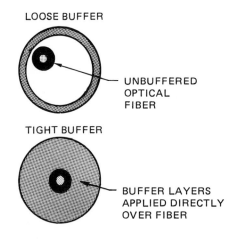

Cable Parameter	Cable Structure	
	Loose Tube	Tight Buffer
Bend Radius	Larger	Smaller
Diameter	Larger	Smaller
Tensile Strength, Installation	Higher	Lower
Impact Resistance	Lower	Higher
Crush Resistance	Lower	Higher
Attenuation Change At Low Temperatures	Lower	Higher

FIGURE 7–3 Loose and tight buffers (Courtesy of Belden Electronic Wire and Cable)

The *tight buffer* has a plastic directly applied over the fiber coating. This construction provides better crush and impact resistance. It does not, however, protect the fiber as well from the stresses of temperature variations. Because the plastic expands and contracts at a different rate than the fiber, contractions caused by variations in temperature can result in loss-producing microbends.

Another advantage to the tight buffer is that it is more flexible and allows tighter turn radii. This advantage can make tight-tube buffers useful for indoor applications where temperature variations are minimal and the ability to make tight turns inside walls is desired.

STRENGTH MEMBERS

Strength members add mechanical strength to the fiber. During and after installation, the strength members handle the tensile stresses applied to the cable so that the fiber is not damaged. The most common strength members are Kevlar aramid yarn, steel, and fiberglass epoxy rods. Kevlar is most commonly used when individual fibers are placed within their own jackets (as in Figure 7–2). Steel and fiberglass members find use in multifiber cables. Steel offers better strength

	PVC	Low Density Polyethylene	Cellular Polyethylene	High Density Polyethylene	Poly-propylene	Poly-urethane	Nylon	Teflon
Oxidation Resistance	E	E	E	E	E	E	E	O
Heat Resistance	G-E	G	G	E	E	G	E	O
Oil Resistance	F	G	G	G-E	F	E	E	O
Low Temperature Flexibility	P-G	G-E	E	E	P	G	G	O
Weather, Sun Resistance	G-E	E	E	E	E	G	E	O
Ozone Resistance	E	E	E	E	E	E	E	E
Abrasion Resistance	F-G	F-G	F	E	F-G	O	E	E
Electrical Properties	F-G	E	E	E	E	P	P	E
Flame Resistance	E	P	P	P	P	F	P	O
Nuclear Radiation Resistance	G	G	G	G	F	G	F-G	P
Water Resistance	E	E	E	E	E	P-G	P-F	E
Acid Resistance	G-E	G-E	G-E	G-E	E	F	P-F	E
Alkali Resistance	G-E	G-E	G-E	G-E	E	F	E	E
Gasoline, Kerosene, Etc. (Aliphatic Hydrocarbons) Resistance	P	P-F	P-F	P-F	P-F	G	G	E
Benzol, Toluol, Etc. (Aromatic Hydrocarbons) Resistance	P-F	P	P	P	P-F	P	G	E
Degreaser Solvents (Halogenated Hydrocarbons) Resistance	P-F	P	P	P	P	P	G	E
Alcohol Resistance	G-E	E	E	E	E	P	P	E

P = poor   F = fair   G = good   E = excellent   O = outstanding
These ratings are based on average performance of general purpose compounds. Any given property can usually be improved by the use of selective compounding.

TABLE 7–1   Properties of jacket materials (Courtesy of Belden Electronic Wire and Cable)

than fiberglass, but it is sometimes undesirable when one wishes to maintain an all-dielectric cable. Steel, for example, attracts lightning, whereas fiberglass does not.

JACKET

The *jacket*, like wire insulation, provides protection from the effects of abrasion, oil, ozone, acids, alkali, solvents, and so forth. The choice of jacket materials depends on the degree of resistance required for different influences and on cost. Table 7–1 compares the relative properties of various popular jacket materials.

When a cable contains several layers of jacketing and protective material, the outer layers are often called the *sheath*. The jacket becomes the layer directly protecting fibers, and the sheath refers to additional layers. This terminology is especially common in the telephone industry.

INDOOR CABLES

Cables for indoor applications include the following:

- Simplex cables
- Duplex cables
- Multifiber cables
- Heavy-, light-, and plenum-duty cables
- Undercarpet cables

Although these categories overlap, they represent the common ways of referring to fibers. Figure 7–4 shows cross sections of several typical cable types.

SIMPLEX CABLES

Simplex cables contain a single fiber. "Simplex" is a term used in electronics to indicate one-way transmission. Since a fiber carries signals in only one direction, from transmitter to receiver, a simplex cable allows only one-way communication.

DUPLEX CABLES

Duplex cables contain two optical fibers. "Duplex" refers to two-way communication. One fiber carries signals in one direction; the other fiber carries signals in the other direction. (Of course, duplex operation is possible with two simplex cables.) Figure 7–5 compares simplex and duplex operations.

In appearance, duplex cables resemble two simplex cables whose jackets have been bonded together, similar to the jacket of common lamp cords. Ripcord constructions, which allow the two cables to be easily separated, are popular.

Duplex cable is used instead of two simplex cables for aesthetics and convenience. It is easier to handle a single duplex cable, there is less chance of the two channels become confused, and the appearance is more pleasing. Remember, the power cord from your lamp is a duplex cable that could easily be two separate wires. Does a single duplex cord in the lamp not make better sense? The same reasoning prevails with fiber-optic cables.

FIGURE 7-4 Indoor cables (Illustration courtesy of AMP Incorporated)

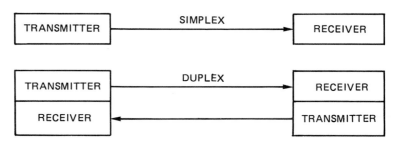

FIGURE 7-5 Simplex and duplex

Optical characteristics

Fiber (μm)	Attenuation (dB/km) 850nm	1,300 nm	Bandwidth (MHz-km) 850 nm	1,300 nm	NA
50/125	<5	<3.5	>400	>400	0.215/0.185
63/125	<6	<4.5	>100	>200	0.305/0.275
85/125	<6	<4.5	>100	>200	0.275/0.245
100/140	<7	<6	>100	>100	0.305/0.275
485/500*	<300 @ 650 nm		>0.5 @ 650 nm		0.50/0.44

Mechanical characteristics

Operating temperature range	0° to 50° C
Storage temperature range	– 20° to 50° C
Minimum bend radius	1.5 in.
Maximum tensile loading	22.5 lb.
Flammability	Passes UL 83

*Plastic fiber

FIGURE 7-6 Undercarpet cable (Illustration courtesy of AMP Incorporated)

MULTIFIBER CABLES

Multifiber cables contain more than two fibers. They allow signals to be distributed throughout a building. Fibers are usually used in pairs, with each fiber carrying signals in opposite directions. A 10-fiber cable, then, permits five duplex circuits.

Multifiber cables often contain several loose-buffer tubes, each containing one or more fibers. The use of several tubes allows identification of fibers by tube, since both tubes and fibers can be color coded. These tubes are stranded around a central strength member of steel or fiberglass. The stranding provides strain relief for the fibers when the cable is bent.

HEAVY-, LIGHT-, AND PLENUM-DUTY CABLES

The first two terms refer to the ruggedness of the cable. Heavy-duty cables can experience more rough handling than light-duty cables, especially during installation. Heavy-duty cables usually have a thicker jacket than light-duty cables.

A plenum is the air space between walls, under structural floors, and above drop ceilings. Plenums are popular places to run signal, power, and telephone lines. Unfortunately, plenums are also an area where fires can easily spread throughout a building. Certain jacket materials give off noxious fumes when burned. The National Electrical Code requires that cables run in plenums must either be enclosed in fireproof conduits or be insulated and jacketed with low-smoke and fire-retardant materials. Plenum cables are those cables whose materials allow them to be used without conduit. Conduits are an additional expense, and they reduce the flexibility of installing and rerouting fibers.

UNDERCARPET CABLE

This fiber does what its name implies. It is run across a floor under carpet squares. A popular use is in the open office. Here work areas are defined by movable walls, partitions, or arrangements of desks and equipment. An important feature is the ability to rearrange or reconfigure the office as needs change.

Another requirement is to bring power, data, and voice lines to equipment. The lines cannot be run through walls, as the power cables in a house are. The walls of an open office are often movable. Wiring alternatives include walls and partitions that contain wiring, underfloor wiring, or undercarpet cables. Copper undercarpet cables for power, voice, and data have become increasing popular since the 1970s. A fiber version is available for voice and data but not for power.

The undercarpet cable shown in cross section in Figure 7–6 contains two optical fibers and two fiber-reinforced plastic strength members enclosed in a polyvinyl chloride jacket. Each fiber is jacketed with polyester. This construction gives a very low profile, only 0.075 inch high, which prevents unsightly bulges in the carpet. The table accompanying the figure shows the fiber sizes available and their optical characteristics.

The cables connect to either wall boxes or floor outlets. The wall boxes allow the transition from intrabuilding wiring to undercarpet. The floor outlets permit the transition from undercarpet cable to the drop cable (simplex or duplex) that runs to the equipment.

One problem with undercarpet cable is making turns without stressing the fibers. The wide cross section of the cable makes turns difficult. Removing the jacket and strength members from around the fibers frees the fibers to bend easily. Unfortunately, the fiber on the outside of the turn must always take a longer path than the fiber on the inside. This unequal path length places differing stresses on the fibers. A turn fitting provides the solution. The fibers lay in tracks in the fitting. The paths of the two tracks have equal lengths that eliminate stresses and allow small turn radii. Additionally, the fitting replaces the strength members and side ramps to provide the crush resistance and low profile required. Figure 7–7 shows the fibers laying in a turn fitting.

OUTDOOR CABLES

Cables used outside of a building must withstand harsher environmental conditions than most indoor cables. Outdoor cables are used in applications such as the following:

FIGURE 7–7 Undercarpet cable and turn fittings (Illustration courtesy of AMP Incorporated)

- Overhead: cables are strung from telephone poles.
- Direct burial: cables are placed directly in a trench dug in the ground and covered.
- Indirect burial: similar to direct burial, but the cable is inside a duct or conduit.
- Submarine: the cable is underwater, including transoceanic applications.

Such cables must obviously be rugged and durable, since they will be exposed to various extremes. Most cables have additional protective sheaths. For example, a layer of steel armoring protects against rodents that might chew through plastic jackets and into the fiber. Other constructions use a gel compound to fill the cables and eliminate air within the cable. Loose-tube buffers, for example, are so filled to prevent water from seeping into the cable, where it will freeze, expand, and damage the cable. The fibers "float" in the gel that will not freeze and damage the fiber.

Most outdoor cables contain many fibers. The strength member is usually a large steel or fiberglass rod in the center, although small steel strands in the outer sheath are also used. Most high-fiber-count cables divide the fibers among several buffer tubes. Figure 7–8 shows the cross section of several typical constructions.

Another approach is ribbon cable. Here, 12 parallel fibers are sandwiched between double-sided adhesive polyester tape. Each ribbon can be stacked with others to make a rectangular array. For example, a stack of 12 ribbons creates an array of 144 fibers. This array is placed in a loose tube, which in turn is covered by two layers of polyethylene. Each polyethylene layer contains 14 steel wires serving as strength members. Depending on the application, additional layers, such as steel armoring, cover the polyethylene. The first commercial fiber-optic sys-

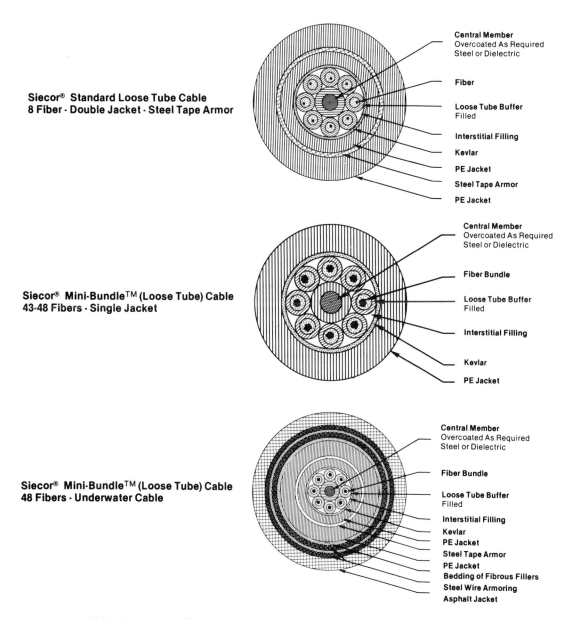

**Siecor® Standard Loose Tube Cable
8 Fiber - Double Jacket - Steel Tape Armor**

- Central Member
  Overcoated As Required
  Steel or Dielectric
- Fiber
- Loose Tube Buffer
  Filled
- Interstitial Filling
- Kevlar
- PE Jacket
- Steel Tape Armor
- PE Jacket

**Siecor® Mini-Bundle™ (Loose Tube) Cable
43-48 Fibers - Single Jacket**

- Central Member
  Overcoated As Required
  Steel or Dielectric
- Fiber Bundle
- Loose Tube Buffer
  Filled
- Interstitial Filling
- Kevlar
- PE Jacket

**Siecor® Mini-Bundle™ (Loose Tube) Cable
48 Fibers - Underwater Cable**

- Central Member
  Overcoated As Required
  Steel or Dielectric
- Fiber Bundle
- Loose Tube Buffer
  Filled
- Interstitial Filling
- Kevlar
- PE Jacket
- Steel Tape Armor
- PE Jacket
- Bedding of Fibrous Fillers
- Steel Wire Armoring
- Asphalt Jacket

FIGURE 7–8 Outdoor multifiber cables (Courtesy of Siecor Corporation)

tem installed by AT&T, in 1977, used a ribbon structure. Figure 7–9 shows such ribbon cable.

In cables containing many fibers, not all the fibers are always used at first. Some are kept as spares to replace fibers that fail in the future. Others are saved for future expansion, as the demands for additional capacity dictate. Installing a

FIGURE 7-9   Ribbon cable (Courtesy of AT & T Bell Laboratories)

cable can be expensive. Having extra fibers in place when they are needed saves future installation costs of additional fibers and cables. It pays to think ahead.

ADDITIONAL CABLE CHARACTERISTICS

Lengths

Cables come reeled in various lengths, typically 1 or 2 km, although lengths of 5 or 6 km are available for single-mode fibers. Long lengths are desirable for long-distance applications, since cables must be spliced end to end over the run. Each splice introduces additional loss into the system. Long cable lengths mean fewer splices and less loss.

Color Coding

Fiber coatings or buffer tubes or both are often color coded to make identification of each fiber easier. In a long-distance link, one must ensure that fiber A in the first cable is spliced to fiber A in the second cable, B to B, C to C, and so forth. Color coding simplifies fiber identification.

Loads

Most cable vendors specify the maximum tensile loads that can be applied to the cable. Two loads are usually specified. The *installation load* is the short-term load that the fiber can withstand during the actual process of installation. This load includes the additional load that is exerted by pulling the fiber through ducts or conduits, around corners, and so forth. The maximum specified installation load limits the length of cable that can be installed at one time, given the particular application. Different applications will offer different installation load conditions. One must carefully plan the installation to avoid overstressing the fiber.

The second load specified is the long-term or *operating load*. During its installed life, the cable cannot withstand loads as heavy as it withstood during installation. The specified operating load is therefore less than the installation load. The operating load is also called the *static load*.

Installation and operational loads are specified in pounds or newtons. Allowable loads, of course, depend on cable construction and intended application. A typical specification for a simplex indoor cable is an installation load of 250 lb (1112 N) and an operational load of 10 lb (44 N).

HYBRID CABLES

Fiber-optic cables sometimes also contain copper conductors, such as twisted pairs. Although the conductors can be used for routine communications,

Optical

Fiber Type: Single Mode

Maximum Attenuation @ 1300nm (dB/km)	1.5
Maximum Dispersion @ 1285nm-1330nm (ps/nm-km)	< 3.5
@ 1550nm	< 20
Zero Dispersion Wavelength, Nominal (nm)	1310
Fiber Mode Diameter (μm):	10±1
Fiber Outer Diameter (μm):	125± 3
Coated Fiber Diameter (μm):	250± 20
Operating Wavelength (nm):	1300
Cut-off Wavelength (nm):	1200± 70

Fiber Type: Graded Index Multimode

Maximum Attenuation (dB/km)	@ 850nm	3.0-4.0
	@ 1300nm	1.0-2.0
Minimum Bandwidth (–3dB) (MHz-km)	@ 850nm	200-800
	@ 1300nm	400-1200
Fiber Core Diameter (μm):		50±3
Fiber Cladding Diameter (μm):		125±3
Coated Fiber Diameter (μm):		245±20
Numerical Aperture:		.235±.015

Physical

	Simplex P/N 01-1020	Duplex P/N 02-1020
OD (inches)	0.12	0.24 × 0.12
Weight (lb/1000 ft)	6	12

Installation

Maximum Pulling Load

Simplex	75 lbf
Duplex	150 lbf

Maximum Operating Load

Simplex	15 lbf
Duplex	20 lbf

Minimum Bend Radius, Dynamic ... 2 inches

Minimum Bend Radius, Static ... 1.2 inches

Temperature Range

Storage	– 40°C to + 80°C
Operating	– 20°C to + 80°C

Mechanical

Crush Resistance ... 228 lbf/in (between parallel plates)

Impact Resistance ... 1000 times @ 1.1 lbf ft (0.5 inch hammer radius)

Flexing ... 6000 times (no cable damage over 1.0 inch radius)

Note: All mechanical and installation specs are calculated on 50/125 multimode optical fiber.

FIGURE 7-10 Typical cable specifications (Illustration courtesy of ITT Electro-Optical Products)

they have two other popular uses. One use is to allow installers to communicate with each other during installation of the fiber, especially with long-distance telephone installation. A technician performing a splice in a manhole must often be in communication with a switching office several miles away. The office contains the test equipment, manned by another technician, that tests the quality of the splice. The other use is to power remote electronic equipment such as repeaters.

UNDERSTANDING CABLE SPECIFICATIONS

Figure 7–10 reproduces a typical data sheet for fiber-optic cable. The sheet lists the specifications most important to the designer planning a fiber-optic system.

Most cable configurations come with various sizes and types of fibers. For example, many fibers have a buffer coating of 250 μm diameter. This coating allows fibers 8/125, 50/125, 62.5, 85/125, or 100/140 to be used. Each of these fibers can further be offered with various attenuations and bandwidths to satisfy the needs of a particular application. Further, a cable using a loose-tube buffer can hold one or several fibers. None of these factors significantly influences cable construction. The same construction can accommodate all these differences easily.

SUMMARY

- Cabling is the packaging of optical fibers for their intended application.
- The main parts of the cable are the buffer, strength member(s), and jacket.
- Cables are available for most application environments.

REVIEW QUESTIONS

1. Name three materials commonly used as strength members.
2. Name two types of cable buffers.
3. Which is greater: installation load or operating load?
4. A main difference between an indoor cable and an outdoor cable is
   A. Fiber bandwidth
   B. Ruggedness and durability
   C. Number of fibers
   D. Attenuation
5. Describe the difference between simplex and duplex cables.
6. Name two uses for copper wires used in a fiber-optic cable.

Chapter 8

Sources

At each end of a fiber-optic link is a transducer, which is simply a device for converting energy from one form to another. The source is an electrooptic transducer—that is, it converts the electrical signal to an optical signal. The detector at the other end is the optoelectronic transducer. It converts optical energy to electrical energy.

The source is either a light-emitting diode (LED) or a laser diode. Both are small semiconductor chips, about the size of a grain of sand, that emit light when current is passed through them. To help you understand the operation of LEDs and lasers, as well as the photodetectors described in Chapter 9, we must first review some of the fundamentals of materials in general and semiconductors in particular.

SOME ATOMIC MATTERS

An atom consists of an inner nucleus around which swirls a cloud of electrons. The electrons circle the nucleus in shells. As shown in Figure 8-1, each shell has a maximum number of electrons in it. The inner shell, K, has a maximum of 2 electrons, L a maximum of 8, M a maximum of 18, and N a maximum of 32. More important for our purposes here, however, is that the outer shell always has a maximum of 8 electrons. This outer shell is called the *valence shell* or *valence band.* The valence band is involved in the chemical bonds that allow elements and compounds to exist.

FIGURE 8-1 Electronic structure of an atom

For an electron to flow as current, it must be free from the valence band to be able to move about in the molecular structure of the material. An electron so freed from the atom is said to be *in the conduction band.* The electron is a free electron because it no longer is bound to the atom. To be available, then, the electron must move from the valence band to the conduction band. How easily this occurs is what separates good conductors from poor conductors.

An atom with only one electron in the valence shell is a very good conductor. Only a small amount of energy is required to promote the electron to the conduction band as a free electron. Copper, gold, and silver, having only one valence electron, are good conductors.

Iron, cobalt, and platinum are examples of elements with eight valence electrons. As a result, they are poor conductors, because an appreciable amount of external energy is required to separate a bound electron. An atom "wishes" to have a filled outer shell. For copper, it easily releases the outer electron, so the next inner shell is filled. For iron, its shell is filled, and it does not wish to release an electron.

A *semiconductor* is a material whose properties fall between those of good conductors like copper and poor conductors like plastics. Separating an electron from the valence shell requires a medium amount of energy. Silicon, for example, has four electrons in its outer shell.

Energy is involved in the movement of an electron between the valence and conduction bands. For an electron to move from the valence band to the conduction band, external energy must be added. An atom that has lost a valence electron develops forces to attract a free electron and complete its shell. When an electron moves from the conduction band to the valence band, it emits energy. The energy released depends on the difference in energy levels between the two bands involved.

Figure 8–2 shows the two energy bands or levels of interest for our understanding of sources and detectors. For an electron to move from one band to another, energy must be either emitted or absorbed. The energy required is the difference between the energy levels of the two bands: $E_1 - E_2$. This difference is the *band gap.* If the band gap is proper—that is, if the difference between the band gap is in a given range—the energy is optical energy, or light. The wavelength of the energy is equal to

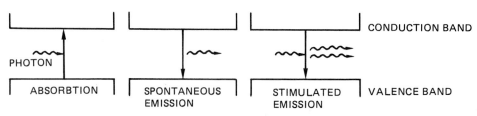

FIGURE 8–2 Absorption and emission

$$\lambda = \frac{hc}{|E_1 - E_2|}$$

where h is Planck's constant (6.625×10^{-34} J-s), c is the velocity of light, and E_1 and E_2 are the energy levels.

In the spontaneous emission shown in Figure 8-2, an electron randomly moves to the lower energy level. An LED works by spontaneous emission. In stimulated emission, a photon of light causes, or stimulates, an electron to change energy levels. A laser operates on the principle of stimulated emission.

In absorption, external energy, such as supplied by a photon of incident light, supplies the energy to allow the electron to move to the higher energy band.

SEMICONDUCTOR PN JUNCTION

The semiconductor pn junction is the basic structure used in the electrooptic devices for fiber optics. Lasers, LEDs, and photodiodes all use the pn junction, as do other semiconductor devices such as diodes and transistors. We will first describe the basic operation of the junction, and then we will describe the operation of LEDs and lasers. Photodiodes are discussed in greater detail in Chapter 9.

As we saw earlier, a silicon atom has four valence electrons. It is these electrons that form the bonds that hold the atoms together in the crystalline structure of the element. For silicon, these bonds are covalent bonds in which the atoms share their electrons in the bond, as shown in Figure 8–3. Here each atom has access to eight valence electrons required for a full shell, four of its own and four from surrounding atoms. All the electrons are taken up in covalent bonds, and none are readily available as free electrons.

The pn junction begins with such a material as silicon. Suppose we add a material with five valence electrons to a silicon crystal. Since only four electrons are needed for the covalent bonds, an electron is left. This electron is a free electron that can move around the structure. It is really a free electron in the conduction band. Because this material has an excess of negatively charged electrons, it is called *n-type* material.

Suppose, on the other hand, we add a material with only three valence electrons to the silicon. One of the covalent bond sites will not be filled by an electron because none is available. This vacancy is called a *hole*. A hole in a semiconductor is a very strange thing because, by definition, it is an absence of something. In appearance, however, it is a charge carrier similar to an electron, but it is positively charged. It appears as a positively charged particle. The material appears to have an excess of positively charged holes, so it is called *p-type* material.

The pn junction of a diode is formed when a semiconductor material is purposely doped with atoms to form p-type and n-type areas separated by a junction. The n-type material has free electrons, and the p-type material has holes. When the p-type and n-type materials are brought together, holes and

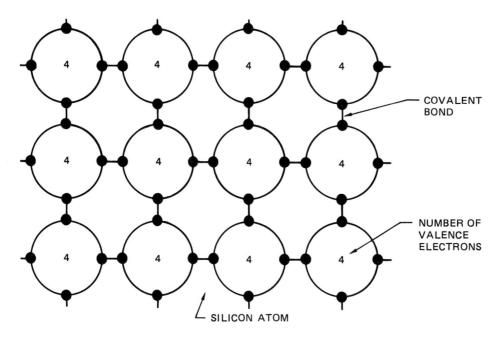

FIGURE 8-3 Covalent bonds in a silicon atom

electrons sweep across the junction and recombine. Recombination means that a free electron "falls" into a hole, moving from the conduction band to the valence band. It becomes part of the covalent bonding structure of the atom. Both the hole and the electron disappear as charge carriers.

Because of the recombination of carriers around the pn junction, no carriers exist in the area. A barrier is formed within this depleted region that prevents further migration of electron and hole carriers across the junction unless additional energy is applied to the material.

When a hole and electron recombine, energy is emitted. Depending on the material, this energy may or may not be light. For silicon, the energy is not light. It is heat in the form of vibrations in the crystal structure.

Lasers and LEDs use elements from Groups III and V of the periodic table. These elements have three and five electrons in their valence shells. If we combine an equal number of atoms with three electrons and an equal number of atoms with five electrons, we will have a situation similar to that of silicon. The atoms will form covalent bonds so that the atoms have filled valence shells. No free carriers exist. To create an n-type material, we combine Group V materials in greater proportion to Group III atoms. Now the structure has free electrons available as carriers. Similarly, a structure having more Group III atoms than Group V atoms will result in holes being available as carriers. Figure 8-4 shows these three combinations using gallium arsenide. Gallium atoms have three valence electrons; arsenic atoms have five valence electrons.

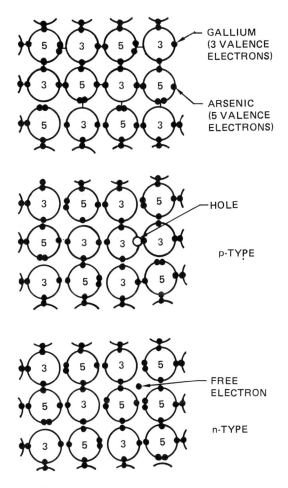

FIGURE 8-4   Holes and free electrons in GaAs

Gallium arsenide has a structure and band gap between the conduction and valence bands such that the recombination of carriers results in emission of light.

An LED is a pn semiconductor that emits light when a *forward bias* is applied (the negative battery terminal connects to the n-type material). In a forward bias, electrons are injected into the n-type material and extracted from the p-type material. (Extracting electrons from p-type material is the same as injecting holes into the material.) Figure 8–5 shows the schematic symbol for an LED and its bias arrangement.

The forward bias causes the electrons and holes to move toward one another and cross the *depletion area* of the junction. Electrons and holes combine, emitting light in the process. For the operation to be maintained, current must continually be supplied to maintain an excess of carriers for recombination. When the drive is removed, recombination of carriers around the junction recreates the depletion area and emission ceases.

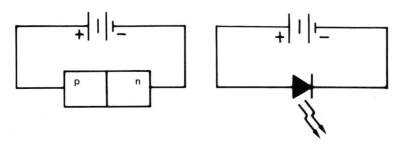

FIGURE 8-5 Light-emitting diode

LEDs

LEDs used in fiber optics are somewhat more complex than the simple device we have just described, although their operation is essentially the same. The complexities arise from the desire to construct a source having characteristics compatible with the needs of a fiber-optic system. Principle among these characteristics are the wavelength and pattern of the emission.

The LED we have described is a *homojunction* device, meaning that the pn junction is formed from a single semiconductor material. Homojunction LEDs are surface emitters giving off light from the edges of the junction as well as its entire planar surface. The result is a low-radiance output whose large pattern is not well suited for use with optical fibers. The problem is that only a very small portion of the light emitted can be coupled into the fiber core.

A *heterojunction* structure solves this problem by confining the carriers to the active area of the chip. A heterojunction is a pn junction formed with materials having similar crystalline structure but different energy levels and refractive indices. The differences confine the carriers and provide a more directional output for the light. The difference in refractive index, for example, can be used to confine and to guide the light in much the same manner that it is confined and guided in an optical fiber. The result is a high-radiance output.

Figure 8–6 shows both a surface-emitting and an edge-emitting LED. The edge-emitting diode uses an active area having a stripe geometry. Because the layers above and below the stripe have different refractive indices, carriers are confined by the waveguide effect produced. (The waveguide effect is the same phenomenon that confines and guides the light in the core of an optical fiber.) The width of the emitting area is controlled by etching an opening in the silicon oxide insulating area and depositing metal in the opening. Current through the active area is restricted to the area below the metal film. The result is a high-radiance elliptical output.

The materials used in the LED influence the wavelengths emitted. LEDs emitting in the first window of 820 to 850 nm are usually gallium aluminum arsenide (GaAlAs). "Window" is a term referring to ranges of wavelengths matched to the properties of the optical fiber. One reason this region was the first exploited in fiber optics is that these devices were better understood, more easily manufactured, more reliable, and less costly. In short, the 820-nm technology was

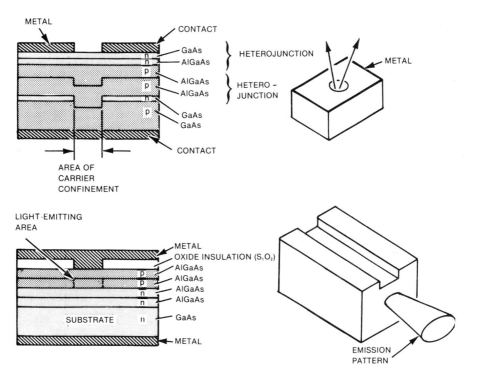

FIGURE 8-6 Surface- and edge-emitting LEDs (Illustration courtesy of AMP Incorporated)

mature, as the 1300-nm technology now is. The movement from 1300 to 1550 nm depends in part on improved technology for sources.

Long-wavelength devices for use at 1300 nm are made of gallium indium arsenide phosphate (GaInAsP) and other combinations of Group III and Group V materials.

LASERS

Laser is an acronym for *light amplification by the stimulated emission of radiation.* Lasers (Figure 8-7) provide stimulated emission rather than the simpler spontaneous emission of LEDs. The main difference between an LED and a laser is that the laser has an optical cavity required for lasing. This cavity, called a *Fabry-Perot cavity,* is formed by cleaving the opposite end of the chip to form highly parallel, reflective mirrorlike finishes.

At low drive currents, the laser acts like an LED and emits light spontaneously. As the current increases, it reaches the threshold level, above which lasing action begins. A laser relies on high current density (many electrons in the small active area of the chip) to provide lasing action. Some of the photons emitted by the spontaneous action are trapped in the Fabry-Perot cavity, reflecting back and

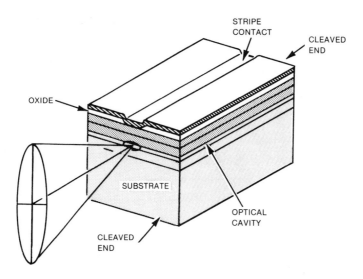

FIGURE 8-7 Semiconductor laser (Illustration courtesy of AMP Incorporated)

forth from end mirror to end mirror. These photons have an energy level equal to the band gap of the laser materials. If one of these photons influences an excited electron, the electron immediately recombines and gives off a photon. Remember that the wavelength of a photon is a measure of its energy. Since the energy of the stimulated photon is equal to the original, stimulating photon, its wavelength is equal to that of the original, stimulating photon. The photon created is a *duplicate of the first photon: It has the same wavelength, phase, and direction of travel.* In other words, the incident photon has stimulated the emission of another photon. Amplification has occurred, and emitted photons have stimulated further emission.

The high drive current in the chip creates population inversion. *Population inversion* is the state in which a high percentage of the atoms move from the ground state to the excited state so that a great number of free electrons and holes exist in the active area around the junction. When population inversion is present, a photon is more likely to stimulate emission than be absorbed. Only above the threshold current does population inversion exist at a level sufficient to allow lasing.

Although some of the photons remain trapped in the cavity, reflecting back and forth and stimulating further emissions, others escape through the two cleaved end faces in an intense beam of light. Since light is coupled into the fiber only from the "front" face, the "rear" face is often coated with a reflective material to reduce the amount of light emitted. Light from the rear face can also be used to monitor the output from the front face. Such monitoring can be used to adjust the drive current to maintain constant power level on the output.

Thus, the laser differs from an LED in that laser light has the following attributes:

- *Nearly monochromatic:* The light emitted has a narrow band of wavelengths. It is nearly *monochromatic*—that is, of a single wavelength. In contrast to the LED, laser light is not continuous across the band of its special width. Several distinct wavelengths are emitted on either side of the central wavelength.
- *Coherent:* The light wavelengths are in phase, rising and falling through the sine-wave cycle at the same time.
- *Highly directional:* The light is emitted in a highly directional pattern with little divergence. *Divergence* is the spreading of a light beam as it travels from its source.

SAFETY

Fiber-optic sources—lasers and high-radiance LEDs, especially—emit intense infrared light invisible to the human eye. Such radiation can permanently damage the retina of the eye and produce visual impairment or blindness. Never look directly into a source or into the end of a fiber energized by a source. Indeed, do not look into a source or fiber unless you are *positive* it is not energized. Remember, you cannot see whether a source or fiber is energized or not. Be careful.

SOURCE CHARACTERISTICS

This section describes some of the main characteristics of interest for sources. In doing so, it compares LEDs and lasers. These characteristics help determine the suitability of an LED or a laser for a given application. Table 8-1 provides a brief comparison of laser and LED characteristics.

OUTPUT POWER

Output power is the optical power emitted at a specified drive current. As shown in Figure 8-8, an LED emits more power than a laser operating below the threshold. Above the lasing threshold, the laser's power increases dramatically

Characteristic	LED	Laser
Output power	Lower	Higher
Speed	Slower	Faster
Output pattern (NA)	Higher	Lower
Spectral width	Wider	Narrower
Single-mode compatibility	No	Yes
Ease of Use	Easier	Harder
Cost	Lower	Higher

TABLE 8-1 Relative characteristics: LED and laser

FIGURE 8-8 Power output versus drive current (Illustration courtesy of AMP Incorporated)

with increases with drive current. In general, the output power of the device is in the following decreasing order: laser, edge-emitting LED, surface-emitting LED.

OUTPUT PATTERN

The output pattern of the light is important in fiber optics. As light leaves the chip, it spreads out. Only a portion actually couples into the fiber. A smaller output pattern allows more light to be coupled into the fiber. A good source should have a small emission diameter and a small NA. The emission diameter defines how large the area of emitted light is. The NA defines at what angles the light is spreading out. If either the emitting diameter or the NA of the source is larger than those of the receiving fiber, some of the optical power will be lost. Figure 8–9 shows typical emission patterns for an LED and a laser.

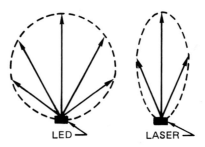

FIGURE 8-9 Emission patterns

The loss of optical power from mismatches in NA and diameter between the source and the core of multimode fiber is as follows: When the diameter of the source is greater than the core diameter of the fiber, the mismatch loss is

$$loss_{dia} = 10 \log_{10} \left(\frac{dia_{fiber}}{dia_{source}} \right)^2$$

No loss occurs when the core diameter of the fiber is larger.

When the NA of the source is larger than the NA of the fiber, the mismatch loss is

$$loss_{NA} = 10 \log_{10} \left(\frac{NA_{fiber}}{NA_{source}} \right)^2$$

No loss occurs when the fiber NA is the larger.

Consider, for example, a source with an output diameter of 100 μm and an NA of 0.30 that is connected to a fiber with a core diameter of 62.5 μm and an NA of 0.29. The losses from diameter and NA mismatches are as follows:

$$loss_{dia} = 10 \log_{10} \left(\frac{62.5}{100} \right)^2$$

$$= 10 \log_{10} (0.390625)$$

$$= -4.1 \text{ dB}$$

$$loss_{NA} = 10 \log_{10} \left(\frac{0.29}{0.30} \right)^2$$

$$= 10 \log_{10} (0.934444)$$

$$= -0.3 \text{ dB}$$

The total mismatch loss is 4.4 dB. If the source output power is 800 μW, only about 290 μW are coupled into the fiber.

A single-mode fiber requires a laser source. The laser provides a small, intense beam of light compatible with the small core of the fiber. The laser's output is elliptical, rather than circular.

SPECTRAL WIDTH

As the discussion of material dispersion in Chapter 5 emphasized, different wavelengths travel through a fiber at different velocities. The dispersion resulting from different velocities of different wavelengths limits bandwidth. Lasers and most LEDs do not emit a single wavelength; they emit a range of wavelengths. This range is know as the *spectral width* of the source. It is measured at 50% of the maximum amplitude of the peak wavelength. For example, if a source has a peak wavelength of 820 nm and a spectral width of 30 nm, its outputs ranges from 805 to 835 nm.

FIGURE 8–10 Typical spectral widths (Illustration courtesy of AMP Incorporated)

At 850 nm, material dispersion can be estimated as 0.1 ns/km for each nanometer of source spectral width. An LED with a 30-nm spectral width results in 3 ns of dispersion over a 1-km run.

Figure 8–10 shows that the spectral width of a laser is substantially narrower than that of an LED. Spectral width of a laser is a few nanometers, but that of an LED is tens of nanometers. As a rule of thumb, spectral width is not important in fiber-optic links running under 100 MHz for a short distance (less than 1 km). Spectral width is especially important in high-speed, long-distance, single-mode systems because the resulting dispersion is the principal limiting factor on system speed. Remember that the practical bandwidth of a single-mode fiber is specified as dispersion in picoseconds per kilometer per *nanometer of source width* (ns/km/nm).

Because laser spectral width is a main limiting characteristic of long-distance, high-speed, single-mode systems, great effort in recent years has been expended to devise and to build reliable single-wavelength laser diodes. Such devices use advanced structures to promote the center wavelength and to suppress others. Two examples of such devices are the *distributed-feedback laser* and the *cleaved-coupled cavity* (C³) laser. The distributed-feedback laser uses an internal grating to limit the photons that can resonate (reflect from the two end walls of the cavity) to one wavelength. Thus, all stimulated emissions have the same wavelength. The C³ laser uses a laser-diode chip separated into two sections by a small gap. Each section is operated separately, and the optical interaction between the two sections purifies the output by suppressing some wavelengths and promoting one wavelength.

SPEED

The source must turn on and off fast enough to meet the bandwidth requirements of the system. Source speed is specified by rise and fall times. Lasers have rise times of less than 1 ns, whereas slower LEDs have rise times of nanoseconds. A rough approximation of bandwidth for a given rise time is

$$BW = \frac{0.35}{t_r}$$

where a rise time in nanoseconds gives a bandwidth in gigahertz. For example, a 1-ns rise time allows 350-MHz operation, and a 5-ns rise time allows 70-MHz operation.

LIFETIME

The expected operating lifetime of a source runs into the millions of hours. Over time, however, the output power decreases because of increasing defects in the device's crystalline structure. The lifetime of the source is typically taken to end when the peak output power is reduced 50% or 3 dB. An LED emitting a peak power of 1 mW, for example, is considered at the end of its lifetime when its peak power becomes 500 μW.

EASE OF USE

Although a laser provides better optical performance than an LED, it is also more expensive, less reliable, and harder to use. Lasers have a shorter expected lifetime than LEDs. They also require more complex transmitter circuits. For example, the output power of a laser can change significantly with temperature. To maintain proper output levels over the required temperature range requires circuitry that detects changes in output and adjusts the drive current accordingly. One method of maintaining these levels is with a photodiode to monitor the light output on the back facet of the laser. The current from the photodiode changes with variations in light output. These variations provide feedback to adjust the laser drive current.

PACKAGING

A central issue in packaging the source for use in fiber optics is to couple as much optical power into the fiber as possible. The optical characteristics of the source must be compatible with the optical characteristics of the fiber. This section looks at many of the techniques used to package the source for fiber-optic systems.

In a surface-emitting LED with a large emitting diameter, only a small portion of this optical energy is coupled into the fiber. In addition, the angles at which light emits tends to be very broad, so that the light spreads out quickly. This fact

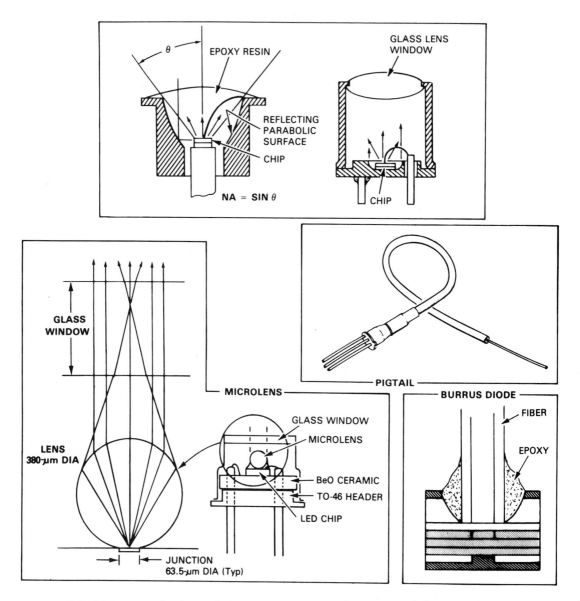

FIGURE 8-11 Packaging techniques attempt to couple maximum light into a fiber (Illustration courtesy of AMP Incorporated)

leads to an important point about sources: The total output power of a chip and the power that is usably coupled into the fiber can vary dramatically.

Although the structure of the semiconductor allows better control of output pattern (as in an edge-emitting diode and laser), source packaging also allows better coupling of light into the fiber. Figure 8-11 shows several techniques used to control the light output.

FIGURE 8-12 Microlensed LED and connector (Courtesy of Hewlett-Packard)

A *microlensed* device uses a small glass bead epoxied directly to the chip. The bead focuses the light in nearly uniform spot on the top of the package. This spot is usually larger than the diameter of fiber. The fiber can be placed anywhere within the spot and receive the same amount of optical energy.

The lens bead can also be placed above the chip to focus the light. In the example shown in Figure 8-12, the lens is part of a mounting to which a fiber-optic connector is attached. The lens directs the light into the fiber in the connector.

Parabolic surfaces and lens caps help to collimate the light, reducing both its emission area and NA. In collimated light, all rays travel parallel to one another, rather than spreading.

Pigtails have a short length of optical fiber as part of the device. A Burrus LED has the fiber directly epoxied into a well etched into the chip. The advantage is to bring the fiber very close to the active region of the device. Other devices have the fiber pigtails aligned very close to the surface of the chip. Pigtails have two advantages. First, placing the fiber close to the chip couples the light into the fiber before the light spreads out. The further away the fiber is from the chip, the more the light spreads out and the less light that enters the fiber. Second, the output power of the pigtailed device is the power at the end of the fiber. This characteristic simplifies system calculations because the emission pattern from the pigtail is more easily specified.

The significance of source packaging can be seen in the offering of one manufacturer. The same LED chip is offered three different packages. The specified output for each package is as follows:

FIGURE 8-13 Laser module (Courtesy of Lytel Incorporated)

- Unpigtailed
  300 μW
- Pigtailed with 100/140 fiber
  250μW
- Pigtailed with 50/125 fiber
  50 μW

The difference in power in the two pigtailed versions results from the greater difficulty of coupling power into the small core and NA of the 50/125 fiber.

Lasers and some high-power edge-emitting LEDs often come packaged in more complex modules, as shown in Figure 8–13. Besides the laser, the modules contain a thermoelectric cooler, heat sink, monitor photodiode, and fiber pigtail. The thermoelectric cooler, a semiconductor device that cools in response to current, helps cool the module, and the heat sink helps dissipate heat. The photodiode allows the output of the laser to be monitored and the drive current to be adjusted to maintain constant levels of output power.

SOURCE EXAMPLES

Table 8–2 provides specifications for typical LEDs; Table 8–3 does the same for laser diodes.

SUMMARY

- The source is the electrooptic transducer in which a forward current results in emission of light.
- An LED emits light spontaneously; a laser uses stimulated emission to achieve higher outputs above the threshold current.

| | DB | JA | WA | YA | WB |

Part No.	λp TYP. (nm)	Δλ TYP. (nm)	pf TYP. (µW)	fC TYP. (MHz)		Mounted fiber					Pack-age	Notes
	I_F=100mA			I_F=100mA$_{DC}$ +20mA$_{P-P}$ −1.5dB from 1MHz	TYPE	Core dia. (µm)	Clad-ding dia. (µm)	NA	Length			
FED086L3DB	865±25	50	300	100	SI	110±5	140±5	0.18±0.02	7.5mm	DB	With φ2.5mm Sleeve Modified TO-18
FED086L4DB	865±25	45	600	45							
FED130L1DB	1300±40	125	100 MIN.	30							
FED130L2DB	1300±40	130	85 MIN.	100							
FED086L1JA	865±25	45	65 MIN.	45	GI	50±3	125±3	0.21±0.02	1m	JA	Pig tail
FED086L2JA	865±25	45	100 MIN.	45							
FED086L3JA	865±25	50	40 MIN.	100							
FED073L1JA	730±5	25	100	30	GI	50±3	125±3	0.21±0.02	1m	JA	Pig tail For WDM
FED078L1JA	780±5	30	100	30							
FED081L1JA	810±5	40	100	40							
FED083L1JA	830±5	40	100	50							
FED088L1JA	880±5	50	100	55							
FED089L1JA	890±5	50	80	55							
FED115L1JA	1150±40	100	30	20	GI	50±3	125±3	0.21±0.02	1m	JA	Pig tail
FED130L1JA	1300±40	125	25	30							
FED130L2JA	1300±40	135	20	100							
FED150L1JA	1480±40	150	15	35							
FED073K1WA	730±15	25	Po 10(mW)	20	−	−	−	−	−	WA	High directivity Modified TO-18
FED086K1WA	865±20	45	Po 10(mW)	40							
FED086K1WB	865±25	45	Po2.5(mW) MIN.	40	−	−	−	−	−	WB	Modified TO-18
FED130K1WB	1300±25	125	Po0.6(mW) MIN.	30							
FED150K1WB	1500±40	170	Po0.35(mW) MIN.	30							
FED086K1YA	865±20	−	Po2.5(mW) MIN.	40	−	−	−	−	−	YA	Open window Modified TO-18
FED130K1YA	1300±25	125	Po0.6(mW) MIN.	30							

TABLE 8-2 LED specifications (Courtesy of Fujitsu Microelectronics, Inc.)

Part No.	TYP. I_{th} (mA)	TYP. V_F (V)	MIN. P_o (mW)	TYP. P_o (mW)	*TYP. P_m (mW)	**TYP. I_m (mA)	TYP. λ_p (nm)	TYP. $\Delta\lambda$ (nm)	***TYP. t_r (ns)	***TYP. t_f (ns)	Package	Notes
	CW	I_F=30mp		$I_F=I_{th}$+30mA			Po=Po(MIN)					
FLD130D4WJ	20	1.2	6	7	2.5	—	1300	2	0.2	0.6	WJ	Modified TO-5
FLD130D4WK	20	1.2	6	7	—	1.8	1300	2	0.2	0.6	WK	With monitor diode
FLD130D3JM	20	1.2	3	3.5	—	1.6	1300	2	0.3	0.8	JM	With monitor diode and multi mode fiber
FLD130D3HM	20	1.2	1.6	2	—	1.0	1300	2	0.3	0.8	HM	With monitor diode and single mode fiber
FLD130D4HM	20	1.2	1.6	2	—	1.0	1300	2	0.2	0.6	HM	
FLD130D3BJ	20	1.2	1	1.3	—	1.3	1300	2	0.3	0.8	BJ	DIL package with monitor diode, peltier cooler and single mode fiber
FLD130D3SJ-A	20	1.2	5	6	2.5	—	1300	2	0.3	0.8	SJ	Chip on carrier
FLD130D4SJ-A	20	1.2	5	6	2.5	—	1300	2	0.2	0.5	SJ	

  * Output power from monitor side
 ** Monitor current of internal photodiode at V_{DR} = 5V
*** $I_{bias} = I_{th}$

TABLE 8-3   Laser specifications (Courtesy of Fujitsu Microelectronics, Inc.)

- A laser has a narrower spectral width than an LED.
- A laser operates faster than an LED.
- Single-mode fibers require a laser.
- LEDs are less expensive and easier to operate than lasers.
- Source packaging is important to efficient coupling of light into a fiber.

REVIEW QUESTIONS

1. What is the purpose of the source?
2. Name the two main types of fiber-optic sources.
3. What is emitted when a free electron recombines with a hole in a semiconductor source?
4. From what phrase is the word *laser* derived?
5. List three packaging methods used to make coupling of light into a fiber more efficient.
6. If a source has a rise time of 7 ns, what is its approximate bandwidth?
7. List three characteristics of laser light that differentiate it from LED light.
8. What relationship between source NA and fiber NA results in NA mismatch loss?

Chapter 9

Detectors

The *detector* performs the opposite function from the source: It converts optical energy to electrical energy. The detector is an optoelectronic transducer. A variety of detector types is available. The most common is the *photodiode,* which produces current in response to incident light. Two types of photodiodes used extensively in fiber optics are the pin photodiode and the avalanche photodiode. This chapter describes photodiode detectors and their characteristics most important to our study of fiber optics.

PHOTODIODE BASICS

Chapter 8 discussed the theory of energy bands in semiconductor material. In moving from the conduction band to the valence band, by recombination of electron-hole pairs, an electron gives up energy. In an LED, this energy is an emitted photon of light with a wavelength determined by the band gap separating the two bands. Emission occurs when current from the external circuit passes through the LED.

With a photodiode, the opposite phenomenon occurs: light falling on the diode creates current in the external circuit. Absorbed photons excite electrons from the valence band to the conduction band, a process known as *intrinsic absorption.* The result is the creation of an electron-hole pair. These carriers, under the influence of the bias voltage applied to the diode, drift through the material and induce a current in the external circuit. For each electron-hole pair thus created, an electron is set flowing as current in the external circuit.

PN PHOTODIODE

The simplest photodiode is the pn photodiode shown in Figure 9–1. Although this type of detector is not widely used in fiber optics, it serves to illustrate the basic ideas of semiconductor photodetection. Other devices—the pin and avalanche photodiodes—are designed to overcome the limitations of the pn diode.

The pn photodiode is a simple pn device. When it is reverse biased (negative battery terminal connected to p-type material), very little current flows. The applied electric field creates a depletion region on either side of the pn junction.

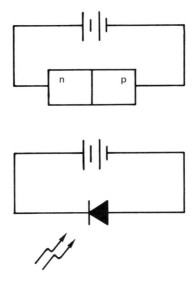

FIGURE 9-1   Pn photodiode

Carriers—free electrons and holes—leave the junction area: in other words, electrons migrate toward the negative terminal of the device (toward the positive terminal of the battery) and holes toward the positive terminal (negative terminal of the battery). Because the depletion region has no carriers, its resistance is very high, and most of the voltage drop occurs across the junction. As a result, electrical forces are high in this region and negligible elsewhere.

An incident photon absorbed by the diode gives a bound electron sufficient energy to move from the valence band to the conduction band, creating a free electron and a hole. If this creation of carriers occurs in the depletion region, the carriers quickly separate and drift rapidly toward their respective regions. This movement sets an electron flowing as current in the external circuit. When the carriers reach the edge of the depletion region, where electrical forces are small, their movement, and hence their external current, ceases.

When electron-hole creation occurs outside of the depletion region, the carriers move slowly toward the depletion region. Many carriers recombine before reaching it. External currents are negligible. Those carriers remaining and reaching the depleted area are swiftly swept across the junction by the large electrical forces in the region to produce an external electrical current. This current, however, is delayed with respect to the absorption of the photon that created the carriers because of the initial slow movement of carriers toward the depletion region. Current, then, will continue to flow after the light is removed. This slow response, due to slow diffusion of carriers, is called *slow tail response*.

Two characteristics of the pn diode make it unsuitable for most fiber-optic applications. First, because the depletion area is a relatively small portion of the diode's total volume, many of the absorbed photons do not result in external current: The created hole and free electrons recombine before they cause external current. The received optical power must be fairly high to generate appreciable current. Second, the slow tail response from slow diffusion makes the diode too slow for medium- and high-speed applications. This slow response limits operations to the kilohertz range.

PIN PHOTODIODE

The structure of the *pin diode* is designed to overcome the deficiencies of its pn counterpart. The depletion region is made as large as possible by the pin structure shown in Figure 9–2. A lightly doped intrinsic layer separates the more heavily doped p-type and n-type materials. *Intrinsic* means that the material is not doped to produce n-type material with free electrons or p-type material with holes. Although the intrinsic layer is actually lightly doped positive, the doping is light enough to allow the layer to be considered intrinsic—that is, neither strongly n-type nor p-type. The name of the diode comes from this layering of materials: *p*ositive, *i*ntrinsic, *n*egative—pin.

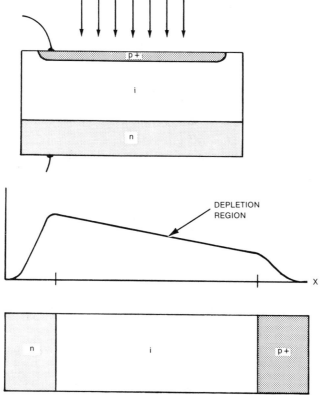

FIGURE 9–2 Pin photodiode (Illustration courtesy of AMP Incorporated)

Since the intrinsic layer has no free carriers, its resistance is high, and electrical forces are strong within it. The resulting depletion region is very large in comparison to the size of the diode. The pin diode works like the pn diode. The large intrinsic layer, however, means that most of the photons are absorbed within the depletion region. The result is improved efficiency in incident photons, creating external current and faster speed. Carriers created within the depletion region are immediately swept by the electric field toward their p or n terminals.

A tradeoff exists in arriving at the best pin photodiode structure. Efficiency of photon-to-carrier conversion requires a thick intrinsic layer to increase the probability of incident photons creating electron-hole pairs in the depletion region. Speed, however, requires a thinner layer to reduce the transit time of the carriers swept across the region. Diode design involves balancing these opposing requirements to achieve the best balance between efficiency and speed.

AVALANCHE PHOTODIODE (APD)

For a pin photodiode, each absorbed photon ideally creates one electron-hole pair, which, in turn, sets one electron flowing in the external circuit. In this sense, we can loosely compare it to an LED. There is basically a one-to-one relationship between photons and carriers and current. Extending this comparison allows us to say that an avalanche photodiode resembles a laser, where the relationship is not one-to-one. In a laser, a few primary carriers result in many emitted photons. In an avalanche photodiode (APD), a few incident photons result in many carriers and appreciable external current.

The structure of the APD, shown in Figure 9–3, creates a very strong electrical field in a portion of the depletion region. *Primary carriers*—the free electrons and holes created by absorbed photons—within this field are accelerated by the field, thereby gaining several electron volts of kinetic energy. A collision of these fast carriers with neutral atoms causes the accelerated carrier to use some of its energy to raise a bound electron from the valence band to the conduction band. A free electron and hole appear. Carriers created in this way, through collision with a primary carrier, are called *secondary carriers*.

This process of creating secondary carriers is known as *collision ionization*. A primary carrier can create several new secondary carriers, and secondary carriers themselves can accelerate and create new carriers. The whole process is called *photomultiplication*, which is a form of gain.

The number of electrons set flowing in the external circuit by each absorbed photon depends on the APD's multiplication factor. Typical multiplication ranges in the tens and hundreds. A multiplication factor of 70 means that, on the average, 70 external electrons flow for each photon. The phrase "on the average" is important. The multiplication factor is an average, a statistical mean. Each primary carrier created by a photon may create more or less secondary carriers and therefore external current.

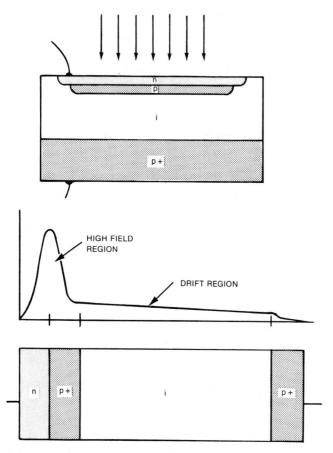

FIGURE 9-3   Avalanche photodiode (Illustration courtesy of AMP Incorporated)

For an APD with a multiplication factor of 70, for example, any given primary carrier may actually create 67 secondary carriers or 76 secondary carriers. This variation is one source of noise that limits the sensitivity of a receiver using an APD. We will discuss noise in fuller detail shortly.

The multiplication factor varies with the bias voltage. Because the accelerating forces must be strong enough to impart energies to the carriers, high bias voltages (several hundred volts in many cases) are required to create the high-field region. At lower voltages, the APD operates like a pin diode and exhibits no internal gain.

The avalanche breakdown voltage of an APD is the voltage at which collision ionization begins. An APD biased above the breakdown point will produce current in the absence of optical power. The voltage itself is sufficient to create carriers and cause collision ionization.

The APD is often biased just below the breakdown point, so any optical power will create a fast response and strong output. The tradeoffs are that dark current (the current resulting from generation of electron-hole pairs even in the absence of absorbed photons) increases with bias voltage, and a high-voltage power supply is needed.

NOISE

The purpose of the detector is to create an electrical current in response to incident photons. It must accept highly attenuated optical energy and produce a current. This current is usually feeble because of the low levels of optical power involved, often only tens or hundreds of nanowatts. Subsequent stages of the receiver amplify and possibly reshape the signal from the detector.

Noise is an ever-present phenomenon that seriously limits the detector's operation. *Noise* is any electrical or optical energy apart from the signal itself. The signal is wanted energy; noise is anything else—that is, unwanted energy. Although noise can and does occur in every part of a communication system, it is of greatest concern in the receiver input. The reason is that the receiver works with very weak signals that have been attenuated during transmission. Although very small compared to the signal levels in most circuits, the noise level is significant in relation to the weak detected signals. The same noise level in a transmitter is usually insignificant because signal levels are very strong in comparison. Indeed, the very limit of the diode's sensitivity is the noise. An optical signal that is too weak cannot be distinguished from the noise. To detect such a signal, we must either reduce the noise level or increase the power level of the signal.

The amplification stages of the receiver amplify both the signal and the noise. Although it is possible to use electronic circuits to filter out some types of noise, it is better to have the signal much stronger than the noise by either having a strong signal level or a low noise level. Having both is even better.

Several types of noise are associated with the photodetector itself and with the receiver. We have already mentioned multiplication noise in an APD, which arises because multiplication varies around a statistical mean. Two other types of noise most important to our understanding of photodiodes and fiber optics are shot noise and thermal noise.

SHOT NOISE

Shot noise arises from the discrete nature of electrons. Current is not a continuous, homogeneous flow. It is the flow of individual, discrete electrons. Keep in mind that a photodiode works because an absorbed photon creates an electron-hole pair that sets an external electron flowing as current. It is a three-step sequence: photon, electron-hole carriers, electron. The arrival and absorption of each photon and the creation of carriers are part of a random process. It is not a perfect homogeneous stream but a series of discrete occurrences. Therefore, the actual current fluctuates, as more or less electron-hole pairs are created in any given moment.

Shot noise exists even when no light falls on the detector. Even without light, a small trickle of current is generated thermally, increasing about 10% for every increase of 1°C. A typical dark current is 25 nA at 25°.

Shot noise is equal to

$$i_{sn}^2 = 2qiB$$

where q is the charge of an electron (1.6×10^{-19} coulomb), i is average current (including dark current and signal current), and B is the receiver bandwidth. The equation shows that shot noise increases with current and with bandwidth. Shot noise is at its minimum when only dark current exists (when i = dark current), and it increases with the current resulting from optical input. A detector with a dark current of 2 nA and operating at a bandwidth of 10 MHz has a shot noise of 80 pA:

$$
\begin{aligned}
i_{sn}^2 &= 2qiB \\
&= (2)(1.6 \times 10^{-19})(2 \times 10^{-9})(10 \times 10^6) \\
&= 6.4 \times 10^{-21} \\
i_{sn} &= 8 \times 10^{-11} \\
&= 80 \text{ pA}
\end{aligned}
$$

THERMAL NOISE

Thermal noise, also called *Johnson noise* or *Nyquist noise*, arises from fluctuations in the load resistance of the detector. The electrons in the resistor are not stationary: Their thermal energy allows them to constantly and randomly move around. At any given instant, the net movement can be toward one electrode or the other, so that randomly varying current exists. This random current will add to and distort the signal current from the photodiode.

Thermal noise is equal to

$$i_{tn}^2 = \frac{4kTB}{R_L}$$

where k is Boltzmann's constant (1.38×10^{-23} J/K), T is absolute temperature (kelvin scale), B is the receiver's bandwidth, and R_L is the load resistance.

Consider a 510-Ω load resistance operating at an absolute temperature of 298 K. Assume a bandwidth of 10 MHz. The thermal noise is

$$
\begin{aligned}
i_{tn}^2 &= \frac{4kTB}{R_L} \\
&= \frac{(4)(1.38 \times 10^{-23})(298)(10 \times 10^6)}{510} \\
&= 3.23 \times 10^{-16} \\
i_{tn} &= 1.79 \times 10^{-8} \\
&= 18 \text{ nA}
\end{aligned}
$$

Thermal and shot noise exist in the receiver independently of the arriving optical power. They result from the very structure of matter. They can be minimized by careful design of devices and circuits, but they cannot be eliminated. Any signal—optical, electrical, or human—must exist in the presence of noise. In the receiver stages following the detector, the signal and noise will be amplified. Therefore, the signal must be appreciably larger than the noise. If the signal power is equal to the noise power, the signal will not even be adequately detected. As a general rule, the optical signal should be twice the noise current to be adequately detected.

SIGNAL-TO-NOISE RATIO

Signal-to-noise ratio (SNR) is a common way of expressing the quality of signals in a system. SNR is simply the ratio of the average signal power to the average noise power from all noise sources.

$$SNR = \frac{S}{N}$$

In decibels, SNR equals

$$SNR = 10 \log_{10} \left(\frac{S}{N} \right)$$

If the signal current is 50 μW and the noise power is 50 nW, the ratio is 1000, or 30 dB.

A large SNR means that the signal is much larger than the noise. The signal power depends on the power of the arriving optical power. Different applications require different SNRs—that is, different levels of "fidelity" or freedom from distortion. The SNR required for a telephone voice channel is less than that required for a television signal, since a fair amount of noise on a telephone line will go unnoticed. We will also accept greater distortion in voices than we will accept in television picture quality. Further, a broadcast-quality television signal, which is the signal produced by the television networks, has a higher SNR than the television signal received in our homes. Why? The broadcast signal itself encounters noise during its transmission. It must, therefore, have a higher SNR so that the signal, after picking up noise during its transmission and reception, still has an SNR high enough to produce a clear, sharp picture.

BIT-ERROR RATE

For digital systems, bit-error rate (BER) usually replaces SNR as a measure of system quality. BER is the ratio of incorrectly transmitted bits to correctly transmitted bits. A ratio of 10^{-9} means that one wrong bit is received for every 1 billion bits transmitted. As with SNR, the required BER varies with the application. Digitally encoded telephone voices have a lower BER requirement than digital computer data, say 10^{-6} versus 10^{-9}. A few faulty bits will not cause noticeable distortion of a voice. A few faulty bits in computer data can cause significant changes in financial data or student grades. They can mean the difference between a program running successfully or crashing.

BER and SNR are related. A better SNR brings a better BER. BER, though, also depends on data-encoding formats and receiver designs. Techniques exist to detect and correct bit errors. We cannot easily calculate the BER from the SNR, because the relationship depends on several factors, including circuit design and bit-error correction schemes. In a given system, for example, an SNR of 22 dB may be required to maintain a BER of 10^{-9}, whereas an SNR of 17 dB brings a BER of 10^{-6}. In another design, however, the 10^{-9} BER might be achieved with an SNR of 18 dB.

DETECTOR CHARACTERISTICS

The detector characteristics of interest here are those that relate most directly to use in a fiber-optic system. These include the device's response to incident optical power and its speed. Since pin diodes are the most commonly used, we concentrate on those.

Responsivity

Responsivity (R) is the ratio of the diode's output current to input optical power and is given in amperes/watt (A/W). Optical power produces current. A pin photodiode typically has a responsivity of around 0.4 to 0.6 A/W. A responsivity of 0.6 A/W means that incident light having 50 μW of power results in 30 μA of current:

$$50 \ \mu W \times 0.6 \ A/W = 30 \ \mu A$$

For an APD, a typical responsivity is 75 A/W. The same 50 μW of optical power now produces 3.75 mA of current:

$$50 \ \mu W \times 75 \ A/W = 3750 \ \mu A = 3.75 \ mA$$

Responsivity varies with wavelength, so it is specified either at the wavelength of maximum responsivity or at a wavelength of interest, such as 850 nm or 1300 nm. Silicon is the most common material used for detectors in the 800- to 900-nm range. Its peak responsivity is 0.7 A/W at 900 nm. At the 850-nm wavelength, responsivity is still near the peak. In a fiber-optic system using plastic fibers, operation is usually at 650 nm in the visible spectrum. Here, the photodiode's responsivity drops to around 0.3 to 0.4 A/W.

Silicon photodiodes are not suitable for the longer wavelengths of 1300 nm and 1550 nm. Materials for long wavelengths are principally germanium (Ge) and indium gallium arsenide (InGaAs). An InGaAs pin photodiode has a fairly broad and flat response curve. It does not peak nearly as sharply as does silicon's. Its

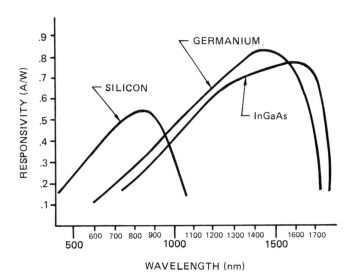

FIGURE 9-4   Responsivity

responsivity of greater than 0.5 A/W from 900 to 1650 nm allows it to be used at both 1300 nm and 1550 nm. Figure 9–4 depicts typical responsivity curves for photodiodes.

Quantum Efficiency (η)

Quantum efficiency is the ratio of primary electron-hole pairs (created by incident photons) to the photons incident on the diode material. It is expressed either as a dimensionless number or as a percentage. A quantum efficiency of 1, or 100%, means that every absorbed photon creates an electron-hole pair. A typical quantum efficiency is 70%, which means only 7 out of overy 10 photons create carriers (electron current). Quantum efficiency applies to primary electrons created by photon absorption not to secondary carriers created by collision ionization.

Quantum efficiency deals with the fundamental efficiency of the diode for converting photons into free electrons. Responsivity can be calculated from quantum efficiency:

$$R = \frac{\eta e \lambda}{hc}$$

where e is the charge of an electron, h is Planck's constant, and c is the velocity of light. Since e, c, and h are constants, responsivity is simply a function of quantum efficiency and wavelength.

Dark Current

As mentioned earlier, *dark current* is the thermally generated current in a diode. It is the lowest level of thermal noise. Dark current increases about 10% for every increase of 1°C. It is much lower in Si photodiodes used at shorter wavelengths than in Ge or InGaAs photodiodes used at longer wavelengths.

Minimum Detectable Power

The minimum power detectable by the detector determines the lowest level of incident optical power that the detector can handle. In simplest terms, it is related to the dark current in the diode, since the dark current will set the lower limit. Other noise sources are factors, including those associated with the diode and those associated with the first stage of the receiver.

The *noise floor* of a pin diode, which tells us the minimum detectable power, is the ratio of noise current to responsivity:

$$\text{Noise floor} = \frac{\text{noise}}{\text{responsivity}}$$

For initial evaluation of a diode, we can use the dark current to estimate the noise floor. Consider a pin diode with $R = 0.5$ $\mu A/\mu W$ and a dark current of 2 nA. The minimum detectable power is

$$\begin{aligned}
\text{Noise floor} &= \frac{2 \times 10^{-12} \text{ A}}{0.5 \text{ A/W}} \\
&= 4 \times 10^{-6} \text{ W} \\
&= 4 \ \mu W
\end{aligned}$$

More precise estimates must include other noise sources, such as thermal and shot noise. As discussed, the noise depends on current, temperature, load resistance, and bandwidth.

Response Time

Response time is the time required for the photodiode to respond to optical inputs and produce external current. As with a source, response time is usually specified as a rise time and a fall time, measured between the 10% and 90% points of amplitude. Rise times range from 0.5 ns to tens of nanoseconds. Rise time is limited to the transit speed of the carriers as they sweep across the depletion region. It is influenced by the bias voltage: higher voltages bring faster rise times. A pin diode, for example, might have a rise time of 5 ns at 15 V and 1 ns at 90 V.

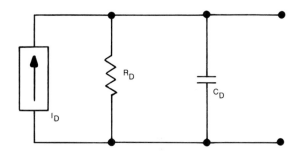

FIGURE 9–5 Electrical model of a pin diode (Illustration courtesy of AMP Incorporated)

The response time of the diode relates to its usable bandwidth. Bandwidth can be approximated from rise time:

$$BW = \frac{0.35}{t_r}$$

The bandwidth, or operating range, of a photodiode can be limited by either its rise time or its *RC* time constant, whichever results in the slower speed or bandwidth. The bandwidth of a circuit limited by the *RC* time constant is

$$BW = \frac{1}{2\pi R_L C_d}$$

where R_L is the load resistance and C_d is the diode capacitance. The rise time of the circuit is

$$t_r = 2.19\ R_L C_d$$

Figure 9–5 shows the equivalent circuit model of a pin diode. It consists of a current source in parallel with a resistance and a capacitance. It appears as a low-pass filter, a resistor-capacitor network that passes low frequencies and attenuates high frequencies. The *cutoff* frequency, which is the frequency that is attenuated 3 dB or 50%, marks the bandwidth. Frequencies higher than cutoff are eliminated.

Diodes for high-speed operation must have capacitances of only a few picofarads or less. The capacitance in a pin diode is mainly the junction capacitance formed at the junctions of p, i, and n layers, as well as any capacitance contributed by the packaging structure and mounting.

Consider a photodiode with a rise time of 1 ns and a capacitance of 2 pF. Its approximate bandwidth, based on rise time, is

$$BW = \frac{0.35}{1\ ns}$$
$$= 0.35\ GHz = 350\ MHz$$

FIGURE 9–6 Effects of bias voltage and temperature on responsivity of an APD (Illustration courtesy of AMP Incorporated)

To ensure that the RC time constant does not lower the bandwidth, we must determine the highest resistor value usable:

$$BW = \frac{1}{2\pi R_L C_d}$$

$$350 \times 10^{-6} \text{ Hz} = \frac{1}{(2)(3.1415927)(2 \times 10^{-12})(R_L)}$$

$$R_L = 227 \ \Omega$$

Thus, a standard 220-Ω resistor works, although in practice a resistor with one quarter of this value is often chosen.

Bias Voltage

Photodiodes require bias voltages ranging from as low as 5 V for some pin diodes to several hundred volts for APDs. Bias voltage significantly affects opera-

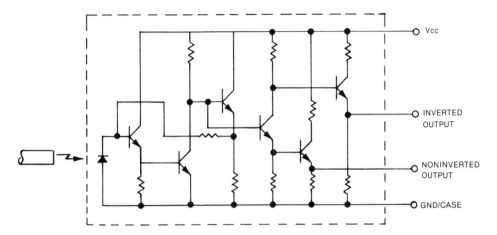

FIGURE 9-7 Integrated detector/preamplifier (Illustration courtesy of AMP Incorporated)

tion, since dark current, responsivity, and response time all increase with bias voltage. APDs are usually biased near their avalanche breakdown point to ensure fast response.

As shown in Figure 9-6, the APD is also sensitive to variations in temperature. The bias voltage required to maintain a given responsivity varies significantly with temperature. The output from an APD becomes erratic unless extensive compensating circuitry is employed in the receiver. The high voltage requirement and temperature sensitivity mean that APDs are chosen only when their responsivity and gain justify their complex application requirements.

INTEGRATED DETECTOR/PREAMPLIFIER

The *integrated detector/preamplifier* (IDP) is an alternative to pin photodiodes. Noise that limits receiver operation can occur between the diode itself and the first receiver stage. The electrical leads of the diode may be sensitive to surrounding EMI and act as an antenna, picking up noise that will be coupled along with the noise into the amplifier.

To reduce these noise sources, the IDP has a *transimpedance amplifier* incorporated on the same semiconductor chip as the photodetector. The transimpedance amplifier performs both amplification and current-to-voltage conversion. The IDP, then, is an integrated circuit having both a photodiode and a transimpedance amplifier. (The amplifier can also be placed in the detector package as a separate chip.) Figure 9-7 shows a schematic of an integrated IDP.

Characteristic specifications for an IDP are similar to those for regular photodetectors. The main difference is that the IDP's output is a voltage, so responsivity is specified in volts/watt (V/W). Since the IDP exhibits gain, typical responsivities are on the order of 40 V/W. For a 50-μW optical input, the output of the IDP would be 2 mV.

| WS | WX | WZ | ST |

Si PIN PHOTODIODES

Part No.	Active area dia. (μm)	*Tolerance MAX. (μm)	η TYP. (%) λ_1=850nm λ_2=830nm	f_C TYP. (MHz) V_R=20V, R_L=50Ω −3dB from 500KHz	I_D TYP. (nA) V_R=20V	C_t TYP. (pF) V_R=20V, f=1MHz	Package	Notes
FID08R42WX	400	200	75 (λ_1)	300	0.1	2	WX	Modified TO-18
FID08R13WX	1000	200	75 (λ_2)	200	2	7	WX	Modified TO-18

Si AVALANCHE PHOTODIODES

Part No.	Active area dia. (μm)	*Tolerance MAX. (μm)	η TYP. (%) λ_1=830nm λ_2=850nm	f_C MIN. (MHz) M=50, R_L=50Ω −1.5dB from 500KHz	V_B (V) I_D=100μA	F TYP. λ_1=830nm, λ_2=850nm M=100 Ipo=2μA, f=30MHz	X TYP.	I_D TYP. (nA) V_R=0.9V_B	Package	Notes
FPD08R32WS	300	85	78 (λ_1)	300	130~180	4.5 (λ_1)	0.33 (λ_1)	0.3	WS	
FPD08R32WX	300	200	70 (λ_2)	300	130~180	4.5 (λ_1)	0.33 (λ_2)	0.3	WX	Modified TO-18

Ge PIN PHOTODIODES

Part No.	Active area dia. (μm)	*Tolerance MAX. (μm)	η TYP. (%) λ=1300nm	f_C TYP. (MHz) R_L=50Ω −3dB from 500KHz	I_D TYP. (μA)	C_t TYP. (pF) f=1MHz	Package	Notes
FID13R13WX	1000	200	75	100 (V_R=10V)	5 (V_R=10V)	30 (V_R=10V)	WX	Modified TO-18
FID13R53WZ	5000	—	73	2.5 (V_R= 3V)	75 (V_R= 3V)	1300 (V_R= 3V)	WZ	Modified TO-8
FID13R13ST	1000	—	70	100 (V_R=10V)	10 (V_R=10V)	30 (V_R=10V)	ST	Chip on carrier

* Tolerance in the alignment between the center of the active area and the center of the package.

TABLE 9-1 Photodiode specifications (Courtesy of Fujitsu Microelectronics, Inc.)

| WS | WT | WX | JT | SR |

Ge AVALANCHE PHOTODIODES

Part No.	Active area dia. (μm)	* Tolerance MAX. (μm)	η TYP. (%) λ=1300nm	f_C TYP. (MHz) M=10, R_L=50Ω –3dB from 500KHz	V_B TYP. (V) I_D=100μA	F TYP.	X TYP. λ=1300nm, M=10 I_{po}=2μA, f=30MHz	i_D TYP. (μA) V_R=0.9V_B	Package	Notes
FPD13R31WS	30	120	80	1000	30	9	0.95	0.1	WS	
FPD13R12WS	100	120	73	700	30	9	0.95	0.3	WS	
FPD13R12WT	100	120	73	700	30	9	0.95	0.3	WT	
FPD13R12WX	100	200	70	700	30	9	0.95	0.3	WX	Modified TO-18
FPD13R32WS	300	120	73	200	30	9	0.95	1.5	WS	
FPD13R32WT	300	120	73	200	30	9	0.95	1.5	WT	
FPD13R12JT	100	—	70	700	30	9	0.95	0.3	JT	** Pig-tail
FPD13R12SR	100	—	73	700	30	9	0.95	0.3	SR	Chip on carrier

InGaAs PIN PHOTODIODES

Part No.	Active area dia. (μm)	* Tolerance MAX. (μm)	η TYP. (%) λ=1300nm	f_C TYP. (MHz) V_R= 5V R_L=50Ω –3dB from 500KHz	I_D TYP. (nA) V_R= 5V	C_t TYP. (pF) V_R= 5V f=1MHz	Package	Notes
FID13S32WS	300	120	76	700	5	4	WS	
FID13S32WX	300	200	75	700	5	4	WX	Modified TO-18
FID13S81WS	80	120	76	1000	1	1	WS	
FID13S81SR	80	—	80	1000	1	0.8	SR	Chip on carrier

* Tolerance in the alignment between the center of the active area and the center of the package.

** With 1m-long GI fiber (core dia. of 50μm and cladding dia. of 125μm)

FIGURE 9-1 (Continued)

PACKAGING

The packaging of the detector is similar to that of the source: TO series cans, pigtailed devices, and microlensed devices predominate. See the discussion of source packaging in Chapter 8 for further details.

As with sources, two main sources of loss in coupling light from a fiber into the detector results from mismatches in diameter and NA: When $dia_{det} < dia_{fiber}$,

$$loss_{dia} = 10 \log_{10} \left(\frac{dia_{det}}{dia_{fiber}} \right)^2$$

When $NA_{det} < NA_{fiber}$,

$$loss_{NA} = 10 \log_{10} \left(\frac{NA_{det}}{NA_{fiber}} \right)^2$$

Since detectors can be easily manufactured with large active diameters and wide angles of view, such mismatches are less common than with sources. Other losses occur from Fresnel reflections and mechanical misalignment between the connector and diode package.

DETECTOR SPECIFICATIONS

Table 9–1 provides specifications for typical pin photodiodes and APDs usable at both short and long wavelengths.

SUMMARY

- The detector is the optoelectronic transducer that converts optical power into current.
- The two most common detectors used in fiber-optic communications are the pin photodiode and avalanche photodiode.
- Responsivity expresses the ratio of output current to optical power.
- An APD provides internal gain, so its responsivity is much higher than that of a pin diode. The APD is also more difficult to apply.
- Noise sets the lower limit of detectable optical power.
- Two types of noise associated with a detector are shot noise and thermal noise.
- SNR and BER are two methods of expressing the "quality" of a signal in a system.
- The response speed of a detector can be limited by its rise time or by its RC time constant.
- An IDP is a detector package containing both a pin photodiode and a transimpedance amplifier.

REVIEW QUESTIONS

1. What is the purpose of the detector?
2. Name the two types of fiber-optic sources.
3. Name the three layers of a pin photodiode. What is the purpose of the middle layer?
4. Name the type of noise that results from current flowing as discrete electrons.
5. Name the type of noise that results from thermal energy in the load resistor.
6. Calculate the current that results from a received optical power of 750 nW and a responsivity of 0.7 A/W.
7. Name two factors that limit the response speed of a detector.
8. Calculate the rise time of a detector circuit having a capacitance of 3 pF and a load resistance of 160 Ω.
9. What distinguishes an IDP from a pin photodiode?
10. Give two reasons why a pin photodiode is used more often than an APD.

Chapter 10

Transmitters and Receivers

In Chapter 1, we saw that a fiber-optic link contains a transmitter, optical cable and receiver. We have looked at fiber-optic cables, sources, and detectors, the electrooptic transducers that provide the bridge between the optical and electronic parts of a fiber-optic system.

BASIC TRANSMITTER CONCEPTS

Figure 10–1 shows a basic block diagram of a transmitter. It contains a driver and a source. The input to the driver is the signal from the equipment being served. The output from the driver is the current required to operate the source.

Most electronic systems operate on standard, well-defined signal levels. Television video signals use a 1-V peak-to-peak level. Digital systems use different standards, depending on the type of logic circuits used in the system. These logic circuits define the levels for the highs and lows that represent the 1s and 0s of digital data.

Currently, the most common digital logic is transistor-transistor logic (TTL). TTL uses 0.5 V for a low signal and 5 V for a high signal. Although TTL is a fast logic, allowing it to be used in many applications, an even faster logic is emitter-

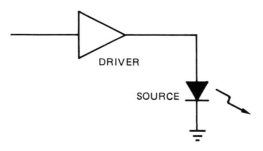

FIGURE 10–1 Basic transmitter block diagram

FIGURE 10–2 TTL and ECL signal levels

coupled logic (ECL). ECL circuits use –1.75 V for low and –0.9 V for high. Figure 10–2 shows the logic levels for TTL and ECL circuits.

Normally, TTL and ECL logics cannot be intermixed; however, special chips allow signals to be converted from one logic level to another. Both TTL and ECL are used in fiber-optic transmitters and receivers. TTL is the more common; ECL is usually only found in high-speed systems—those above 50 or 100 Mbps.

Another very popular logic family is complementary metal-oxide semiconductor (CMOS). It is rapidly becoming a popular replacement for TTL because of its very low power consumption. Many CMOS circuits use the same voltage levels as TTL for their signals.

The drive circuit of the transmitter must accept these logic levels. It then provides the output current to drive the source. For example, it may convert the 0.5 V and 5 V of TTL into 0 mA and 50 mA to turn the source on and off.

An additional function of the transmitter is to produce the proper modulation code.

MODULATION CODES

A string of digital highs and lows is often not suitable for transmission over any appreciable distance. Whereas digital circuits use simple high and low pulses to represent 1s and 0s of binary data, a more complex format is often used to transmit digital signals between electronic systems. A modulation code is a method of encoding digital data for transmission.

Earlier, we saw that the pulse represents a binary 1, and the absence of a pulse represents a binary 0. In a TTL system, a 5-V high represents a 1; a 0.5-V low represents a 0. Thus, there is a one-to-one correspondence between a high or low voltage and a binary 1 or 0. With the modulation codes described here, this correspondence does not always exist.

Figure 10–3 shows several popular modulation codes. Each bit of data must occur with its bit period, which is defined by the clock. The *clock* is a steady string of pulses that provides basic system timing. As shown in the table in Figure 10–3, some codes are self-clocking, and others are not. A self-clocking code means that the clock information is contained within the code. In a non-self-clocking system, this timing information is not present.

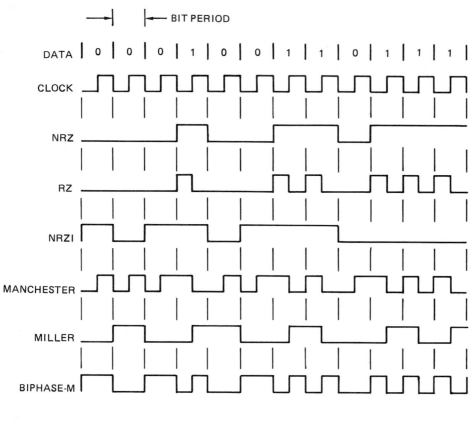

FIGURE 10-3 Modulation codes

FORMAT	SYMBOLS PER BIT	SELF-CLOCKING	DUTY FACTOR RANGE (%)
NRZ	1	NO	0 - 100
RZ	2	NO	0 - 50
NRZI	1	NO	0 - 100
MANCHESTER (BIPHASE-L)	2	YES	50
MILLER	1	YES	33 - 67
BIPHASE-M (BIFREQUENCY)	2	YES	50

Clock information is important to the receiver. One purpose of a receiver is to rebuild signals to their original state as they were presented to the transmitter. To do so, the receiver must know the timing information. There are three alternatives:

1. The transmitted information can also contain clock information; in other words, the modulation code is self-clocking.
2. The clock or timing information must be transmitted on another line. This, of course, adds to system complexity by increasing the number of lines from transmitter to receiver. In a long-distance system, such costs may be considerable.
3. The receiver may provide its own timing and not rely on clock signals from the transmitter.

NRZ Code

The *NRZ (nonreturn-to-zero) code* is similar to "normal" digital data. The signal is high for a 1 and low for a 0. For a string of 1s, the signal remains high. For a string of 0s, it remains low. Thus, the level changes only when the data level changes.

RZ Code

The *RZ (return-to-zero) code* remains low for 0s. For a binary 1, the level goes high for one half of a bit period and then returns low for the remainder. For each 1 of data, the level goes high and returns low within each bit period. For a string of three 1s, for example, the level goes high for each 1 and returns to low.

NRZI Code

In an *NRZI (nonreturn-to-zero, inverted) code,* a 0 is represented by a change in level, and a 1 is represented by no change in level. Thus, the level will go from high to low or from low to high for each 0. It will remain at its present level for each 1. An important thing to notice here is that there is no firm relationship between 1s and 0s of data and the highs and lows of the code. A binary 1 can be represented by either a high or a low, as can a binary 0.

Manchester Code

A *Manchester code* uses a level transition in the middle of each bit period. For a binary 1, the first half of the period is high, and the second half is low. For a binary 0, the first half is low, and the second half is high.

Miller Code

In the *Miller code,* each 1 is encoded by a level transition in the middle of the bit period. A 0 is represented either by no change in level following a 1 or by a change at the beginning of the bit period following a 0.

Biphase-M Code

In the *biphase-M code*, each bit period begins with a change of level. For a 1, an additional transition occurs in midperiod. For a 0, no additional change occurs. Thus, a 1 is at both high and low during the bit period. A 0 is either high or low, but not both, during the entire bit period.

DATA RATE AND SIGNAL RATE

A close inspection of the modulation codes in Figure 10–3 shows an important aspect of signal transmission. The highs and lows may be changing faster than the 1s and 0s of binary data. In the figure, there are 12 data bits (000100110111). The modulation codes, however, use more symbols to represent these bits. A symbol is a high or a low pulse in the modulation code. The Manchester code uses two symbols, both a high and a low, for each binary bit. It always requires twice as many symbols as there are bits to be transmitted.

When we describe the speed of a system, it becomes important to distinguish between data rate and signal rate. *Data rate* is the number of *data bits* transmitted in bits per second. A system may operate at 10 Mbps. The *signal* rate (or speed) is the number of *symbols* transmitted per second. Signal speed is expressed in baud. The signal speed (or baud rate) and the data rate (or bit rate) may or may not be the same, depending on the modulation code used.

For NRZ data, which uses one symbol per bit, the rates are the same. A 10 Mbps requires a 10 Mbaud capacity. For Manchester-encoded data, which uses two symbols per bit, the baud rate is twice that of the bit rate. To transmit a 10-Mbps data stream requires a transmission link having a bandwidth of 20 Mbaud. The baud rate, or number of symbols to be transmitted, is the true indication of a system's signaling speed. This fact is true for all transmission systems, fiber optic or otherwise.

DUTY CYCLE

Duty cycle refers to the ratio of high to low symbols in the encoded string. A duty cycle of 0 means that all symbols are low. A duty cycle of 100 means that all symbols are high. A duty cycle of 50 means that there are an equal number of highs and lows. Duty cycle expresses the relationship between peak and average power levels arriving at the receiver. For a duty cycle of 50%, the average power is half the peak power. Above a 50% duty cycle, the average power increases in relation to peak power. Below 50%, it declines. Duty cycle becomes important in the receiver, and we will delay further discussion until later in this chapter.

TRANSMITTER OUTPUT POWER

The output power of the transmitter is of great concern. The power of many transmitters is specified for the power coupled into a given fiber. Thus for a given transmitter design, the typical output power might be as follows:

Fiber	Power
50/125 μm, 0.21 NA	−16 dBm
62.5/125 μm, 0.29 NA	−12 dBm
85/125 μm, 0.26 NA	−10 dBm
100/140 μm, 0.30 NA	−8 dBm

The power coupled increases with core diameter and NA. The tradeoff, of course, is that the smaller cores and NA usually indicate fibers with lower attenuation and higher bandwidths.

If the transmitter power output is not specified as listed, the power coupled into the fiber must be calculated. To do so, we must know the source output power, output diameter and NA, and the fiber core diameter and NA, and the expected connector loss. We will look at an example of this calculation in the next chapter.

BASIC RECEIVER CONCEPTS

Figure 10-4 shows a basic diagram of a receiver. It contains the detector, amplifier, and output section. The amplifier amplifies the attenuated signal from the detector. The output section can perform many functions:

- Separation of the clock and data
- Pulse reshaping and retiming
- Level shifting to ensure compatibility—TTL, ECL, and so forth—with the external circuits.
- Gain control to maintain constant amplification levels in response to variations in received optical power and variations in receiver operation from temperature or voltage changes.

Because the receiver deals with the highly attentuated light signals, it can be considered the principal component around which the design or selection of a fiber-optic system revolves. It is in the photodetector and first stage of amplification that the signal being transmitted is at its weakest and most distorted. It is reasonable to say that this is the central part of the link. Decisions affecting other

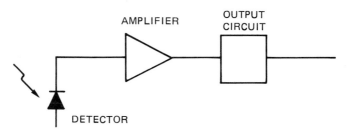

FIGURE 10-4 Basic receiver block diagram

parts of the link are decided with the receiver in mind. Decisions about the modulation codes in the transmitter are decided, at least in part, by the requirements of the receiver.

Receiver Sensitivity

Receiver sensitivity specifies the weakest optical signal that can be received. The minimum signal that can be received depends on the noise floor of the receiver front end. The last chapter described the limitations on the minimum optical power the detector can receive. The minimum level for the receiver is basically the same, except that the influence of amplifier noise must also be included. From the noise floor, the lowest *practical* optical power to the receiver depends on the SNR or BER requirements of the system.

Sensitivity can be expressed in microwatts or dBm. Expressing a sensitivity as 1 μW or –30 dBm is saying the same thing. With a packaged receiver, the sensitivity may be specified as an absolute minimum (the noise floor) or with respect to a given level of performance such as a BER level.

Dynamic Range

Dynamic range is the difference between the minimum and maximum acceptable power levels. The minimum level is set by the sensitivity and is limited by the detector. The maximum level is set by either the detector or the amplifier. Power levels above the maximum saturate the receiver or distort the signal. The received optical power must be maintained below this maximum.

If a receiver has a minimum optical power requirement of –30 dBm and a maximum requirement of –10 dBm, its dynamic range is 20 dB. It can receive optical inputs between 1 and 100 μW.

Figure 10–5 shows the relationship between BER, data rate, and received power for a typical receiver. As we would expect, BER improves as the received power becomes larger. On the other hand, the required power for a given BER increases significantly as the data rate becomes higher. A BER of 10^{-9} requires an optical input of about –35 dBm (310 nW) for a 25-Mbps data rate and about –32 dBm (620 nW) for a 60-Mbps rate. The received power must be 3 dB, or about 50%, greater for the higher data rate.

AMPLIFIER

Two classical designs used in fiber-optic receivers use a low-impedance amplifier and a transimpedance amplifier. Examples are shown in Figure 10–6.

The bandwidth of the low-impedance amplifier is determined by the *RC* time constant of the circuit:

$$BW = \frac{1}{2\pi RC}$$

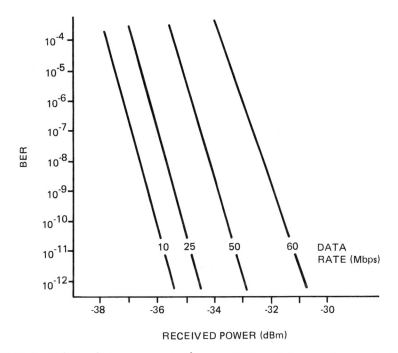

FIGURE 10-5   Relation between received power, BER, and data rate for a typical optical receiver

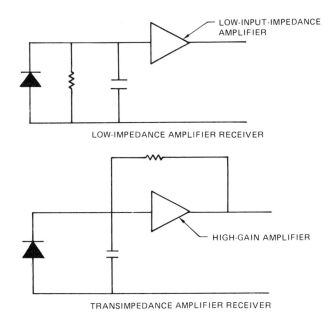

FIGURE 10-6   Low-impedance and transimpedance circuit designs

This equation is the same as for a photodiode whose response time is limited by the time constant. From the discussion of noise and response time for a photodiode, recall that bandwidth and resistor noise decrease with increasing resistance. Therefore, receiver sensitivity can be increased by increasing the load resistor but at the expense of bandwidth. The resulting high-impedance amplifier has a much smaller bandwidth but greater sensitivity.

The bandwidth of a receiver with a transimpedance amplifier is affected by the gain of the amplifier:

$$BW = \frac{g}{2\pi RC}$$

where g is the open-loop gain. To achieve a bandwidth comparable to the low-impedance circuit, the transimpedance circuit must have a much lower value for the resistor across the amplifier.

Representative sensitivities for the different receiver designs operating at 100 Mbps and a BER of 10^{-9} are

- low impedance
  –33 dBm
- high impedance
  –40 dBm
- transimpedance
  –36 dBm

DUTY CYCLE IN THE RECEIVER

The reason for the concern on duty cycle in the modulation codes is that some receiver designs put restrictions on the duty cycle. A receiver distinguishes between high and low pulses by maintaining a reference threshold level. A signal level above the threshold is seen as a high or 1; a signal level below the threshold is seen as a low or 0.

In a restricted-duty-circle receiver, a duty cycle far removed from 50% will increase the chances of a level being misinterpreted. A high level is seen as a low level, or a low level is misinterpreted as a high level. As the duty cycle departs from the 50% ideal, the misinterpretations of signal levels lower the BER.

Figure 10–7 shows why such errors occur in a receiver with a restricted duty cycle receiving NRZ code. The receiver sets the threshold level based on the average optical power being received. The top part of the figure shows how the threshold level depends on the duty cycle. For a 50% duty cycle, the threshold is set midway between the high and low signal levels, since the average power is midway between signal levels. The difference between the threshold and a low signal level is the same as the difference between the threshold and the high signal level.

At a 20% duty cycle, where low levels significantly outnumber high levels, the average received power is much lower. The threshold shifts much closer to the low signal level. At an 80% duty cycle, the average received power is much higher,

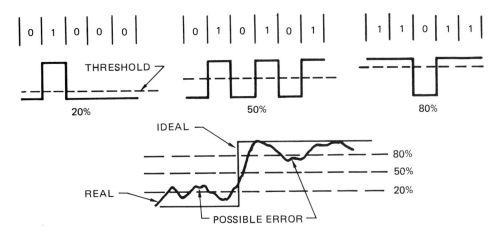

FIGURE 10-7 Errors and duty cycle in a restricted-duty-cycle receiver

and the threshold shifts closer to the high signal level. This shifting results from the average power being received.

This shifting of threshold level would cause no problems in an ideal, noiseless receiver. But receivers are neither perfect nor noiseless. Signal levels not only vary somewhat, but the signals contain noise. The bottom half of the figure shows how noise distorts the pulse. If the signal level crosses the threshold at the wrong point, it is misinterpreted as the wrong level. A 0 can cross the low threshold at a 20% duty cycle and be seen as a 1. A 1 can cross the threshold at an 80% duty cycle and be seen as a 0. In either case, a bit error results.

There are two ways to get around such errors. The first is to maintain a duty cycle close to 50%. Manchester and biphase-M codes, by definition, always have a 50% duty cycle, so they satisfy the requirement. Their drawback is that they require a channel bandwidth of twice the data rate. They also increase the complexity of the transmitter somewhat.

The second method of avoiding bit errors is to design a receiver that maintains the threshold without drift. The reference threshold is always midway between high and low signal levels. One way to do this is by a dc-coupled receiver, which is designed to operate with arbitrary data streams. The receiver is edge-sensing, meaning that it is sensitive to changes in level and not to the levels themselves. This type of receiver reacts only to pulse transitions.

Since the dc-coupled receiver is easy to design and build, why not use it all the time? The tradeoff is receiver sensitivity. It often requires 6 to 8 dB higher optical power levels. This power penalty must be paid by lower data rates, shorter transmission distances, or lower error requirements. Even so, it may often be the best approach in many applications.

In ac coupling, which is one approach used in restricted-duty-cycle receivers, any dc levels are removed from the signal. Only the time-varying portions are left. This stripping of dc levels is often done capacitively, since a capacitor blocks dc and passes ac. The receivers in Figure 10-5 are shown dc coupled to keep the representation simple. An ac-coupled design would place a capacitor between the photodetector and the amplifier.

TRANSCEIVERS AND REPEATERS

Transceivers and repeaters are two important components in many fiber-optic applications. They are simple enough in concept, and we need only comment briefly on each. A *transceiver* is a transmitter and receiver packaged together to allow both transmission and reception from either station. A *repeater* is a receiver driving a transmitter. The repeater is used to boost signals when the transmission distance is so great that the signal will be too highly attenuated before it reaches the receiver. The repeater accepts the signal, amplifies and reshapes it, and sends it on its way by feeding the rebuilt signal to a transmitter.

One advantage of digital transmission is that it uses regenerative repeaters that not only amplify a signal but reshape it to its original form as well. Any pulse distortions from dispersion or other causes are removed. Analog signals use nonregenerative repeaters, which amplify the signal, including any noise or distortion. Analog signals cannot be easily reshaped, because the repeater does not know what the original signal looked like. For a digital signal, it does know.

TRANSMITTER AND RECEIVER PACKAGING

The physical packaging of a transmitter or receiver can be an important aspect in many applications. It can be a small module placed on a printed circuit board, an independent printed circuit board plugged into equipment, or a stand-alone unit. Part of the choice lies in the complexity of the design, and part in the needs of the application. Figure 10–8 shows board-mounting transmitter and receiver modules.

TRANSMITTER AND RECEIVER SPECIFICATIONS

Many of the characteristics for specifying a transmitter or receiver are those of concern with any electronic circuit. These include power supply voltages, storage and operating temperature ranges, required input and output voltage levels (which indicate TTL or ECL compatibility), propagation delays, and so forth. There are also important specifications relating to the transmitter or receiver in a fiber-optic system. We have already become acquainted with each of these.

Transmitter
- Peak output power
- Data rate/bandwidth
- Operating wavelength
- Source spectral width
- Duty-cycle restrictions
- Rise and fall times

FIGURE 10-8 Transmitter/receiver modules (Illustration courtesy of AMP Incorporated)

Receiver
- Data rate/bandwidth
- Sensitivity
- Dynamic range
- Operating wavelength
- Duty-cycle restrictions

Table 10–1 supplies the specifications for a 100-Mbps fiber-optic transmitter-receiver pair. From the specifications, the following points become apparent:

- The power out of the transmitter into the fiber depends significantly on the type of fiber. Power into the fiber varies from –8 dBm for a 100/140 fiber to –16 dBm for a 50/125 fiber.
- Optical power into the receiver does not depend on fiber type.
- The peak output varies over a 4-dB range between minimum and maximum.
- Duty-cycle restrictions apply to the receiver, not to the transmitter. At 10 Mbps, a 50% duty cycle is required. At 100 Mbps, the duty cycle ranges between 40% and 60%.
- The specified data rate is for NRZ modulation codes. For codes such as Manchester or biphase-M, which require two symbols per data bit, the data rate becomes 50 Mbps, and the signal rate is 100 Mbaud.

TRANSMITTER					
Parameter	**Condition**	**Min**	**Typ**	**Max**	**Units**
Data Rate		0		100	Mbps
Optical Output (Peak)	50/125 fiber 0.21 NA $V_I = -0.9$ V	-20		-16	dBm
	62.5/125 fiber 0.29 NA $V_I = -0.9$ V	-16		-12	dBm
	85/125 fiber 0.26 NA $V_I = -0.9$ V	-14		-10	dBm
	100/140 fiber 0.3 NA $V_I = -1.6$ V	-12		-8	dBm
Optical Wavelength		805	815	825	nm
Spectral Width	FWHM			50	nm
Duty Cycle	$B = 0 - 100$ Mbps	0		100	%
Rise Time (Output)	20%–80%	2.5		4.0	ns
Fall Time (Output)	20%–80%	2.5		4.0	ns
Supply Voltage (V_{EE})		-5.72	-5.2	-4.68	V
Operating Temperature		0	25	70	C

TABLE 10-1 Transmitter and receiver specifications (Illustration courtesy of AMP Incorporated)

SUMMARY

- A transmitter includes the source and its drive circuit.
- A receiver includes the detector, amplification circuits, and output circuits.
- Modulation codes are the formats by which digital bits are encoded for transmission.
- For some modulation codes, data rate and baud rate are the same. For others, they are not the same.
- Some modulation codes are self-clocking.
- Duty cycle expresses the ratio between high levels and low levels.
- The receiver sensitivity sets the lowest acceptable optical power that can be received.
- The dynamic range of the receiver is the difference between the minimum and maximum optical power that a receiver can handle.

REVIEW QUESTIONS

1. Sketch a block diagram of a simple transmitter.
2. A Manchester-encoded transmission has a data rate of 200 Mbps. What is its baud rate?

RECEIVER					
Parameter	**Condition**	**Min**	**Typ**	**Max**	**Units**
Data Rate		10		100	Mbps
Optical Input (Peak)	50/125 fiber 0.21NA BW = 100 Mbps BER = 10^{-12}	−31		−6	dBm
	62.5/125 fiber 0.29 NA BW = 100 Mbps BER = 10^{-12}	−31		−6	dBm
	85/125 fiber 0.26 NA BW = 100 Mbps BER = 10^{-12}	−31		−6	dBm
	100/140 fiber 0.3 NA BW = 100 Mbps BER = 10^{-12}	−31		−6	dBm
Optical Wavelength in		800		900	nm
Duty Cycle	BW = 100 Mbps	40		60	%
	BW = 10 Mbps	50		50	%
Rise Time (Output)	20%–80%	1		3.5	ns
Fall Time (Output)	20%–80%	1		3.5	ns
Supply Voltage (V_{CC})		4.75	5	5.25	V
Supply Voltage (V_{EE})		−5.72	−5.2	−4.68	V
Operating Temperature		0		70	C

TABLE 10-1   (Continued)

3. Name two examples of self-clocking modulation codes.
4. Sketch an NRZ pattern for 1111001. Be sure to show each bit period.
5. Will a typical transmitter couple more power into a 50/125-μm fiber or an 85/125-μm fiber? Why?
6. A receiver has a sensitivity of −30 dBm and a maximum receivable power of −10 dBm. What is its dynamic range?
7. Name two of the commonest types of receiver circuits.
8. Sketch a block diagram of a repeater.
9. If a receiver has a sensitivity of −28 dBm, what is the minimum optical power it can receive?

Chapter 11

Connectors and Splices

Interconnection of the various components of a fiber-optic system is a vital part of system performance. Connection by splices and connectors couples light from one component to another with as little loss of optical power as possible. Throughout a link, a fiber must be connected to sources, detectors, and other fibers. This chapter describes the basic considerations involved in fiber-optic interconnections and describes several examples of actual connectors and splices. To simplify the discussion, we emphasize connecting one fiber to another.

By popular usage, a *connector* is a disconnectable device used to connect a fiber to a source, detector, or another fiber. It is designed to be easily connected and disconnected many times. A *splice* is a device used to connect one fiber to another permanently. Even so, some vendors do offer disconnectable splices that are not permanent and that can be disconnected for repairs or rearrangement of circuits.

The requirements for a fiber-optic connection and a wire connection are very different. Two copper conductors can be joined directly by solder or by connectors that have been crimped or soldered to the wires. The purpose is to create intimate contact between the mated halves to maintain a low-resistance path across the junction. Connectors are simple, easy to attach, reliable, and essentially free of loss.

The key to a fiber-optic interconnection is precise alignment of the mated fiber cores (or spots in a single-mode fibers) so that nearly all the light is coupled from one fiber across the junction into the other fiber. Contact between the fibers is not even mandatory. The demands of precise alignment on small fibers create a challenge to the designer of the connector or splice.

THE NEED FOR CONNECTORS AND SPLICES

Connectors and splices are required for many reasons. Although these reasons may be fairly obvious since they mirror the same needs with metallic conductors, we will briefly review them here.

In a long link, fibers may have to be spliced end to end because cable manufacturers offer cables in limited lengths—typically 1 to 6 km. If the cable comes in 6-km lengths, a 30-km span requires four splices (as well as connections to the transmitter and receiver). In other cases, it may not be practical to even pull a 6-km length of cable through ducts during installation. More moderate lengths may be easier to install.

Connectors or splices may be required at building entrances, wiring closets, couplers, and other intermediate points between transmitter and receiver. These allow, for example, transitions between outdoor and indoor cables, rearrangement of circuits, and the division of optical power from one fiber into several fibers.

Finally, fibers must be connected to the source in the transmitter and the detector in the receivers.

Dividing a fiber-optic system into several subsystems connected together by splices and connectors also simplifies system selection, installation, and maintenance. Components may be selected from different vendors. They may be installed by different vendors or contractors. For example, a building contractor can wire a building with fiber-optic cable, a manufacturer of computer terminals can build fiber-optic transmitters and receivers in the equipment, and the telephone company can bring a fiber-optic telephone line into the building. All these various parts are then linked together with connectors. Maintenance of the system is simplified when a faulty or outdated part can be disconnected and a new or updated part installed. Faster transmitters and receivers, for instance, can be installed without disturbing the fiber.

CONNECTOR REQUIREMENTS

The following is a list of desirable features for a fiber-optic connector or splice:

- *Low loss:* The connector or splice should cause little loss of optical power across the junction.
- *Easy installation:* The connector or splice should be easily and rapidly installed without need for extensive special tools or training.
- *Repeatability:* A connector should be able to be connected and disconnected many times without changes in loss.
- *Consistency:* There should be no variation in loss; loss should be consistent whenever a connector is applied to a fiber.
- *Economical:* The connector or splice should be inexpensive, both in itself and in special application tooling.

It can be very difficult to design a connector that meets all the requirements. A low-loss connector may be more expensive than a high-loss connector, or it may require relatively expensive application tooling. The lowest losses are desirable, but the other factors clearly influence the selection of the connector or splice.

In general, loss requirements for splices and connectors are as follows:

- < 0.5 dB for telecommunication and other splices in long-haul systems.
- 0.5 to 1 dB for connectors used in intrabuilding systems, such as local area networks or automated factories.
- 1 to 3 dB for connectors and splices used in applications where such higher losses are acceptable and low cost becomes more important than low loss. Such applications often use plastic fibers.

CAUSES OF LOSS IN AN INTERCONNECTION

Three different types of factors cause loss in fiber-optic interconnections:

1. Intrinsic or fiber-related factors are those caused by variations in the fiber itself.
2. Extrinsic or connector-related factors are those contributed by the connector itself.
3. System factors are those contributed by the system.

INTRINSIC FACTORS

In joining two fibers to one another, we would like to assume that the two fibers are identical. Usually, however, they are not. The fiber manufacturing process allows fibers to be made only within certain tolerances. From the nominal specification, a fiber will vary within stated limits. Figure 11–1 shows schematically

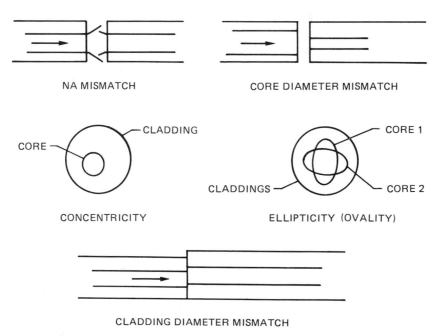

FIGURE 11-1 Factors of intrinsic interconnection loss

TYPE OF VARIATION	TOLERANCE
Core diameter (50 μm)	$\pm 3\ \mu$m
Cladding diameter (125 μm)	$\pm 3\ \mu$m
NA (0.260)	± 0.015
Concentricity	$\leq 3\ \mu$m
Core Ovality	≥ 0.98
Cladding Ovality	≥ 0.98

TABLE 11-1 Typical tolerances that influence intrinsic loss

the most important variations in tolerances that cause loss. The following paragraphs discuss them in greater detail.

NA-mismatch loss occurs when the NA of the transmitting fiber is larger than that of the receiving fiber. *Core-diameter-mismatch loss* occurs when the core or diameter of the transmitting fiber is larger than that of the receiving fiber. *Cladding-diameter-mismatch loss* occurs when the claddings of the two fibers differ, since the cores will not longer align.

Concentricity loss occurs because the core may not be perfectly centered in the cladding. Ideally, the geometric axes of the core and cladding should coincide. The concentricity tolerance is the distance between the core center and the cladding center.

Ellipticity (or *ovality*) *loss* occurs because the core or cladding may be elliptical rather than circular. Notice that the alignment of two elliptical cores will vary, depending on how the fibers are brought together. At one joining, they may be at the extremes of ellipticity so that maximum loss occurs. At the next joining, one fiber may become rotated in relation to the other, so their elliptical axes are the same. No loss then occurs. The ellipticity or ovality tolerance of the core and cladding equals the minimum diameter divided by the maximum diameter.

These variations, shown in Figure 11-1, exist in any fiber. The fiber manufacturer controls these variations by manufacturing fiber to tight, exacting tolerances. In the past few years, manufacturing techniques have improved sufficiently that fiber tolerances are much tighter. For example, a 125-μm-diameter fiber once had a tolerance of $\pm 5\ \mu$m, so the actual fiber diameter was in a range of 120 to 130 μm. Two fibers at the extreme of tolerance mismatch present a loss of 0.6 dB. Today, the normal tolerance is $\pm 3\ \mu$m, resulting in a range of 122 to 128 μm and a maximum loss of 0.4 dB. A tolerance of $\pm 1\ \mu$m reduces the loss to 0.1 dB.

Remember that these losses are maximums and will not be experienced in most cases since the probability of joining two fibers at the extreme diameter mismatches is small.

Table 11-1 shows typical tolerances for today's off-the-shelf fibers.

EXTRINSIC FACTORS

Connectors and splices also contribute loss to the joint. When two fibers are not perfectly aligned on their center axes, loss occurs even if there is no intrinsic

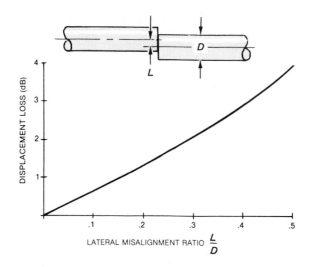

FIGURE 11–2   Loss from lateral displacement (Illustration courtesy of AMP Incorporated)

variation in the fibers. The loss results from the difficulty of manufacturing a device to the exacting tolerances required. As we will see, many different alignment mechanisms have been devised for joining two fibers.

The four main causes of loss that a connector or splice must control are

1. Lateral displacement
2. End separation
3. Angular misalignment
4. Surface roughness

The following discussion assumes perfect fibers, without tolerance variation. The curves presented are general ones for multimode fibers. As discussed later in this chapter, modal conditions in the fiber also influence loss across the junction.

Lateral Displacement

A connector should align the fibers on their center axes. When one fiber's axis does not coincide with that of the other, loss occurs. As shown in the graph in Figure 11–2, the amount of displacement depends on the ratio of the lateral offset to the fiber diameter. Thus the acceptable offset becomes less as the fiber diameter becomes smaller. A displacement of 10% of the core axis yields a loss of about 0.5 dB. For a fiber with a 50-μm core, a 10% displacement is 5 μm, which, in turn, means that each connector half must contribute only a 2.5 μm.

These dimensions would double for a fiber with a 100-μm core: a 10% displacement would allow 10 μm or 5 μm for each connector half. Obviously, then, control of lateral displacement becomes more difficult for smaller-

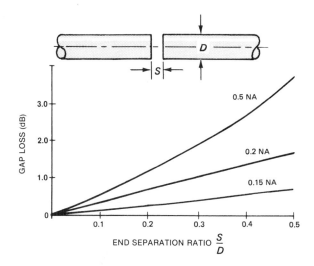

FIGURE 11-3 Loss from end separation (Illustration courtesy of AMP Incorporated)

diameter fibers. Connector manufacturers attempt to limit displacement to less than 5% of the core diameter.

End Separation

Two fibers separated by a small gap experience two types of loss, as shown in Figure 11-3. First, is *Fresnel reflection loss* caused by the difference in refractive indices of the two fibers and the intervening gap, which is usually air. Fresnel reflection occurs at both the exit from the first fiber and at the entrance to the second fiber. For glass fibers separated by air, loss from Fresnel reflections is about 0.34 dB. Fresnel losses can be greatly reduced by using an index-matching fluid between the fibers. *Index-matching fluid* is an optically transparent liquid or gel having a refractive index the same as or very near that of the fibers.

The second type of loss for multimode fibers results from the failure of high-order modes to cross the gap and enter the core of the second fiber. High-order rays exiting the first fiber will miss the acceptance cone of the second fiber. Light exiting the first fiber spreads out conically. The degree of separation loss, then, depends on fiber NA. A fiber with a high NA cannot tolerate as much separation to maintain the same loss as can a fiber with a lower NA.

Ideally, the fibers should butt to eliminate loss from end separation. In most splices, the fibers do butt. In a separable connector, a very small gap is sometimes desirable to prevent the ends from rubbing together and causing abrasion during mating. Fibers brought together with too much force, as may be done by over-tightening coupling nuts on a connector, may even be fractured. Therefore, some connectors are designed to maintain a very small gap between fibers, and others use spring pressure to bring the fiber ends gently together without the danger of damage.

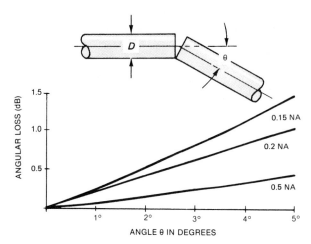

FIGURE 11-4   Loss from angular misalignment (Illustration courtesy of AMP Incorporated)

Angular Misalignment

The ends of mated fibers should be perpendicular to the fiber axes and perpendicular to each other during engagement. The losses graphed in Figure 11-4 result when one fiber is cocked in relation to the other. Again, the degree of loss also depends on fiber NA. Notice, however, that the influence of NA is opposite that for end separation. Here larger NA can tolerate greater angular misalignment to maintain the same loss as for a lower NA.

A properly applied connector controls angular misalignment rather easily, so its loss contribution is not as great as with lateral displacement. This control includes having the fiber face perpendicular to the fiber axis, which is a function of proper fiber cleaving or polishing, and preventing the connector from cocking during engagement.

Surface Finish

The fiber face must be smooth and free of defects such as hackles, burrs, and fractures. Irregularities from a rough surface disrupt the geometrical patterns of light rays and deflect them so they will not enter the second fiber.

SYSTEM-RELATED FACTORS

The loss at a fiber-to-fiber joint depends not only on the losses contributed by the connector and fiber but on system-related factors as well. Chapter 5 discussed how modal conditions in a fiber change with length until the fiber reaches equilibrium mode distribution (EMD). Initially, a fiber may be overfilled or fully filled, with light being carried both in the cladding and in high-order modes. Over distance, these modes will be stripped away. At EMD, a graded-index fiber has a reduced NA and a reduced active area of the core carrying the light.

Consider a connection close to the source. The fiber on the transmitting side of the connection may be overfilled. Much of the light in the cladding and high-order modes will not enter the second fiber, although it was present at the junction. This same light, however, would not have been present in the fiber at EMD, so it would also not have been lost at the interconnection point.

Next consider the receiving side of the fiber. Some of the light will spill over the junction into cladding and high-order modes. If we were to measure the power from a short length of fiber, these modes would still be present. But these modes will be lost over distance, so their presence is misleading.

Similar effects will be seen if the connection point is far from the source, where the fiber has reached EMD. Since the active area of a graded-index fiber has been reduced, lateral misalignment will not affect loss as much, particularly if the receiving fiber is short. Again, light will couple into cladding and high-order modes. These modes will be lost in a long receiving fiber.

Thus, the performance of a connector depends on modal conditions and the connector's position in the system (since modal conditions vary along the length of the fiber). In evaluating a fiber-optic connector or splice, we must know conditions on both the launch (transmitter) side and the receive (receiver) side of the connection. Four different conditions exist:

1. Short launch, short receive
2. Short launch, long receive
3. Long launch, short receive
4. Long launch, long receive

With all other things held constant, connector performance depends on the launch and receive conditions. For a series of tests done on one connector, for example, losses under long-launch conditions were in the 0.4- to 0.5-dB range. Under short-launch conditions, the losses were in the 1.3- to 1.4-dB range. Thus, a difference of nearly 1 dB occurred simply as a result of the launch conditions used in the test.

INSERTION LOSS

Insertion loss is the method for specifying the performance of a connector or splice. Power through a length of fiber is measured. Next, the fiber is cut in the center and the connector or splice is applied. Power at the end of the fiber is again measured. Insertion loss is

$$loss_{IL} = 10 \log_{10} \left(\frac{P_1}{P_2} \right)$$

where P_2 is the initial measured power and P_1 is the measured power after the connector is applied.

This testing method, by using a single fiber, minimizes the influence of fiber variation on loss. By rejoining the cut fiber, we make the mated halves essentially identical. The only fiber variations present are eccentricity and ellipticity. NA and diameter variations are eliminated. The purpose is to evaluate connector performance independently of fiber-related variations.

Launch and receive conditions still affect the measured insertion loss. Therefore, these conditions must be known when comparing insertion loss specifications for various connectors. Differences in loss specifications may result from differences in test conditions. In addition, the application in which the connector will be used must also be considered. It is best to evaluate a connector that has been tested under conditions similar to those in which it will be used.

ADDITIONAL LOSSES IN AN INTERCONNECTION

The discussion of intrinsic fiber loss stated that loss occurs when the transmitting fiber has a core diameter or NA smaller than the core diameter or NA of the receiving fiber. This discussion assumed the connection was between two fibers of the same type, differing only in variations in tolerance. In this case, these variations may effect the performance of the connector, but the losses from the mismatches are not usually calculated in general loss budgets. If, however, two different types of fibers are connected, diameter- and NA-mismatch losses may be significant and must be accounted for.

When the NA of the transmitting fiber is greater than that of the receiving fiber, the NA-mismatch loss is

$$loss_{NA} = 10 \log_{10} \left(\frac{NA_r}{NA_t} \right)^2$$

Similarly, when the diameter of the transmitting fiber is greater than that of the receiving fiber, the diameter-mismatch loss is

$$loss_{dia} = 10 \log_{10} \left(\frac{dia_r}{dia_t} \right)^2$$

Assume, for example, a transmitting fiber with a core diameter of 100 μm and an NA of 0.30 and a receiving fiber with a diameter of 50 μm and an NA of 0.20. Such a connection could be experienced in connecting a local area network to outside telephone lines. Loss from NA-mismatch is

$$loss_{NA} = 10 \log_{10} \left(\frac{NA_r}{NA_t} \right)^2$$

$$= 10 \log_{10} \left(\frac{0.20}{0.30} \right)^2$$

$$= -3.5 \text{ dB}$$

Loss from diameter mismatch is

$$\text{loss}_{\text{dia}} = 10 \log_{10} \left(\frac{\text{dia}_r}{\text{dia}_t} \right)^2$$

$$= 10 \log_{10} \left(\frac{50}{100} \right)^2$$

$$= -6.0 \text{ dB}$$

The total loss is 9.5 dB.

LOSS IN SINGLE-MODE FIBERS

Connectors and splices for single-mode fibers must also provide a high degree of alignment. In many cases, the percentage of misalignment permitted for a single-mode connection is greater than for its multimode counterpart. Because of the small size of the fiber core, however, the actual dimensional tolerances for the connector or splice remain as tight or tighter. For example, a gap of 10 times the core diameter results in a loss of 0.4 dB in a single-mode fiber.

Mismatches in mode field diameter (spot size) are the greatest cause of intrinsic, fiber-related loss. Most manufacturers limit spot size deviations to 10%, which results in attenuation increases of under 0.1 dB in 99% of the cases where fibers from the same manufacturer are connected. Since different manufacturers have different values for mode field diameter, greater losses may be expected when mating connectors from different manufacturers.

ALIGNMENT MECHANISMS AND CONNECTOR EXAMPLES

Many different mechanisms have been used to achieve the high degree of alignment required in a connector or splice. This section looks at examples of actual connectors and splices and the means they use to achieve alignment. You will thus see both the alignment mechanism in concept and its application in a practical product.

In most cases, splices are simpler than connectors. The permanence of splices allows a connection device that has fewer component parts. A fusion splice, in fact, has no parts in its simplest form.

FUSION SPLICES

A *fusion splice* uses an electric arc to weld two glass fibers together. After the fibers have been prepared by the scribe-and-break method, they are placed on a V-groove of an alignment block of the splice machine. Micromanipulators bring the fibers together both axially and laterally. Alignment can be observed visually through the splicer's microscope, as shown in Figure 11–5, or by monitoring light

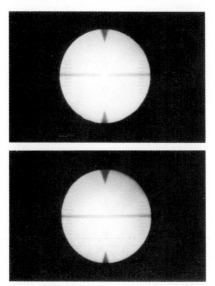

FIGURE 11-5   View of fusion splice through microscope (Courtesy of Siecor Corporation)

across the junction. Once the fibers are aligned, an electric arc is applied; the localized heating softens the butted fiber ends and fuses them. Surface tension during fusion tends to align the fiber axes and minimize lateral displacement. The finished splice is protectively encased in epoxy or heat-shrink tubing.

ELASTOMERIC SPLICE

The *elastomeric splice* consists of two elastomeric inserts held within a quartz glass sleeve (Figure 11-6). One insert has a flat surface, and the other has a molded precision V-groove. The triangular space formed by the flat and the V-groove is smaller than the fiber to be inserted. Each insert contains a tapered entrance to ease the insertion of the fiber. When optical fibers are slid into the inserts, the elastomer distorts equally on each of the three sides to align the fibers on their center axes. Two fibers of different diameters distort the elastomer differently and therefore remain concentrically aligned on their center axes.

An index-matching gel in the V-groove not only reduces mismatch losses, but it also serves as a lubricant to aid fiber insertion. Fibers are bonded in the end guides with an ultraviolet (UV) -cured adhesive. A optional plastic housing offers further protection and strain relief.

During assembly of the splice, power through the junction can be measured so one fiber can be rotated in relation to another to minimize loss.

FOUR-ROD SPLICE

The alignment mechanism for a *four-rod splice* is a four-rod glass array with slight bends at each end. The interstitial space formed by the rods is several times

FIGURE 11-6 Elastomeric splice (Courtesy of GTE Fiber Optic Products)

larger than the fibers. The bends at each end force the fibers to rest in one of the grooves formed by two rods. This arrangement provides close lateral and angular alignments and shock and vibration immunity. The splice uses index-matching fluid to reduce loss and a UV-cured adhesive to bond the fibers.

Silicon strain reliefs protect the fiber from mechanical stress, whereas a stainless steel housing crimped to the silicon sleeves provides protection and mechanical support (Figure 11–7).

Cross Section of Four Rod Glass Alignment Element

FIGURE 11–7 Four-rod splice (Courtesy of Connector Division, TRW Electronics Components Group)

The same four-rod alignment approach has also been used for connectors (Figure 11–8). One connector half holds the glass-rod array in a plastic slug. The other half houses a spring-loaded piston that protects the freestanding fiber. When the connector halves mate, the piston is pushed back to bare the fiber, which enters the alignment rods in the other half and aligns with the other fiber.

A symmetrical scissor clamp in each connector half secures the cable jacket and strength members within the connector shell and prevents movement of the fiber within the connector.

RIBBON CABLE ARRAY CONNECTOR

A *ribbon cable array connector* is used to splice ribbon cables of the sort discussed in Chapter 7 and shown in Figure 7–9. The connector consists of two array halves, two negative chips with metal backing, and two spring clips (Figure 11–9). Each array half, consisting of two positive silicon chips into which grooves have been etched, is bonded to the fibers at the factory before the cable is shipped. Grinding and polishing the array prepares the fiber ends simultaneously.

In the field, the connector is assembled by aligning the two array halves within the two negative chips. Each negative chip overlaps both array halves to

Cross Section of Four Rod
Glass Alignment Element

Scissor Clamp

Cable

Strain
Relief

FIGURE 11–8 Four-rod connector (Courtesy of Connector Division, TRW Electronics Components Group)

align the fibers. Spring clips hold the assembly together. After assembly, index-matching fluid is injected.

Several array connectors can be stacked in a single assembly. One type of cable, for example, has 144 fibers in 12 ribbons. Thus, 12 array connectors are required. Although the connector was not designed for repeated connection and disconnection, the connector can be reentered for testing or fiber rearrangement.

SMA CONNECTORS

The fiber-optic *SMA connector* is very popular with the military and test equipment industry. It is one of the few connectors that is a "standard" component in which offerings from different vendors are interchangeable and inter-matable. Standardization applies to the mating dimensions of the connector and not the internal details of the connector and alignment mechanism. The acronym SMA derives from the size of the coupling nut, which is the same as that of an SMA rf (radio frequency) coaxial connector.

The SMA connector in Figure 11–10 consists of a metal body assembly with a precision hole bored in its tip. A stripped fiber, held within a primary alignment ferrule, is inserted into the back of the body assembly and secured with epoxy. Strength members, if present in the cable, are trapped and crimped between the body assembly and eyelet to provide mechanical strength.

SIDE VIEW AND PARTIAL CROSS SECTION OF ARRAY CONNECTOR

CROSS SECTION OF ARRAY CONNECTOR

FIGURE 11–9   Ribbon cable array connection (Courtesy of AT & T Bell Laboratories)

The connector is attached to a metal polishing bushing, the excess fiber is removed, and the assembly is polished to the proper length and optical finish.

The connectors mate in a straight sleeve bushing, whose precise internal diameter provides a friction fit. Additional alignment within the sleeve depends on the specific connector. Some use an alignment sleeve that attaches to the step-down on the nose of the connector. The sleeve provides a tight friction fit within the bushing to center the connector. The nose of the mating connector also enters the sleeve during mating. Another version of the SMA connector uses

FIGURE 11-10   SMA connectors (Illustration courtesy of AMP Incorporated)

a resilient connector tip. The tip contains small dimples around its circumference that compress the tip and center the fiber within the sleeve.

Heat-shrink tubing applied over the cable and connector provides strain relief.

PLASTIC-FERRULE SIMPLEX CONNECTOR

In a *plastic-ferrule simplex connector*, fibers are held within plastic ferrules that mate in a bushing that tapers inward to a common cylinder bore

FERRULE

FERRULES COMPRESSED INTO BUSHING

FIGURE 11–11   Plastic-ferrule simplex connector (Illustration courtesy of AMP Incorporated)

(Figure 11–11). The fiber hole is precision molded in the center of the ferrule face. The tapering of the bushing and ferrule nose provide additional alignment during mating, and the shoulder on the ferrule provides a positive stop controlling the depth of the ferrule into the bushing.

The connector assembly is mechanically held together with epoxy and a crimp. The epoxy bonds the fiber within the ferrule. A double crimp applied to the metal retaining assembly serves two purposes: (1) It attaches the metal retaining assembly to the ferrule; (2) it provides mechanical attachment of cable strength members, which are trapped between the ferrule and metal retaining assembly.

DNP CONNECTOR

For plastic fibers, connector cost may outweigh performance. Since a connector loss of 2 or 3 dB is still only a small percentage of the total link loss, a very-low-cost connector is appropriate for use with plastic fibers in links with

FIGURE 11-12 DNP connector (Illustration courtesy of AMP Incorporated)

lengths under 30 m and speeds under 30 Mbaud. The plastic DNP (for dry, no-polish) connector in Figure 11-12 requires neither the epoxy nor the crimp associated with the other connectors discussed. The plugs, which hold the fiber, have split tips with opposing V-grooves. The tip opens just enough to allow a slightly oversized fiber to pass through while the grooves center the fiber. A retention clip holds the fiber jacket inside the plug. The clip has interior and exterior barbs that engage the fiber jacket and the inside of the plug. Two plugs mate within a straight-sleeve receptacle. Dimples on the outside of the plug serve as compression points for centering the fiber. The need for optional polishing may be relieved by trimming the fiber at the tip of the plug with a cutting instrument resembling a hand soldering iron with a sharp blade for a tip.

FIGURE 11-13   Active device mount (Illustration courtesy of AMP Incorporated)

CONNECTION TO SOURCES AND DETECTORS

Connectors such as the SMA or simplex connectors are easily adapted for connection to sources and detectors. The bushing used to mate the connector halves is modified to house the source or detector package. The active device is inserted into the back end of the mount, and the connector threads onto the front end. Figure 11-13 shows an example that uses an alignment ring to hold and center the active device.

If the source or detector is pigtailed, a connector is applied to the pigtail fiber. The connection is simply fiber to fiber. A simplex or SMA connector, for example, can be attached to the end of the pigtail.

FIBER PREPARATION

Proper preparation of the fiber end face is critical to any fiber-optic connection. The two main features to be checked for proper preparation in end finish are *perpendicularity* and *end finish*. The end face ideally should be perfectly square to the fiber and practically should be within 1° or 2° of perpendicular. Any divergence beyond 2° increases losses unacceptably. The fiber face should have a smooth, mirrorlike finish free of blemishes, hackles, burrs, and other defects.

The two common methods used to produce correct end finishes are the *scribe-and-break method* (also called the score-and-break method) and the *polish method*. The scribe-and-break method is used mostly with splices, whereas the polish method is more commonly used with connectors.

Whichever method is used, the fibers must first be bared. Jackets, buffer tubes, and other outer layers are removed with wire strippers, cutting pliers, utility knives, and other common tools. Metal strength members are removed with wire cutters, and scissors are used to cut Kevlar strength members.

The plastic buffer coating attached to the cladding can be removed chemically or mechanically. Chemical removal involves soaking the fiber end for about 2 min in a paint stripper or other solvent and then wiping the fiber clean with a soft tissue. Mechanical removal is done with a high-quality wire stripper. Care must be taken not to nick or damage the cladding. The bared fiber can be cleaned with isopropyl alcohol or some other suitable cleaner.

The scribe-and-break method uses a fiber held under slight pressure. A cutting tool with a hard, sharp blade, such as diamond, sapphire, or tungsten carbide, scribes a small nick across the cladding. The blade can be pulled across a stationary fiber, or a fiber can be pulled across a stationary blade. After scribing, the pressure is increased by pulling, which forces the flaw to propagate across the face of the fiber.

Properly done, the cleave produces a perpendicular, mirrorlike finish. Improperly done, the cleave results in hackles and lips, as shown in Figure 11–14, that make the cleave unacceptable. The technique must then be repeated.

Scribe-and-break cleaving can be done by hand or by tools that range from relatively inexpensive hand tools to elaborate automated bench tools. Any technique or tool is capable of good cleaves; the trick is consistent finishes time and time again. In general, the less costly approaches require more skill and training of technicians making the cleave.

Most connectors use polishing to achieve the proper end finish after the fiber has been partially or completely assembled in the connector. For example, for the connectors in Figure 11–11, polishing is the final step of assembly. Most connectors use some sort of polishing fixture, which has a large polishing surface to ensure a perpendicular finish. A smaller-faced fixture runs the risk that the assembly will be cocked or tilted during hand polishing.

Polishing is done in two or more steps with repeatedly finer polishing grits, typically down to 1 μm or 0.3 μm. The simplex connector in Figure 11–11 uses a 5-μm-grit lapping film followed by a 1-μm film. The SMA connector in Figure 11–10 uses 15-, 5-, and 1-μm films. Polishing is done with a figure-8 motion. Wet polishing is recommended, preferably under running water, which not only lubricates and cools the fiber, but also flushes polish remnants away. The connector and fiber face should be cleaned before switching to a finer polishing material.

Polishing has a second function: It grinds the connector tip to a precise dimension. This dimension controls the depth that the connector tip and fiber extend into the bushing that holds the two connectors. It thereby controls the gap between mated fibers. If the tip dimension is too long, the mated fibers may be damaged when they are brought together. If the dimension is too short, the gap may be large enough to produce unacceptable losses. The first polishing steps grind the connector tip and fiber to the correct dimension. The final step polishes the fiber face to a mirrorlike finish.

As with a cleaved fiber, the polished fiber should be inspected under a microscope. Small scratches on the fiber face are usually acceptable, as are small pits on the outside rim of the cladding. Large scratches, pits in the core region, and fractures indicate unacceptable end finishes. Some poor finishes, such as scratches, can be remedied with additional polishing with 1-μm, or finer, film. Fractures and pits usually mean a new connector must be installed. Figure 11–15 shows examples of acceptable and unacceptable polishes.

CONNECTOR ASSEMBLY EXAMPLE

To demonstrate the practical assembly of a connector, we reproduce in this section the assembly instructions for a duplex connector from AMP Incorporated

FIGURE 11-14 Good and bad cleaves of an optical glass fiber (Courtesy of GTE Fiber Optic Products)

(Figure 11–16). The connector uses the same plastic ferrule as the simplex connector discussed earlier. It is, however, a duplex connector holding two fibers within a plug kit. A common use is for duplex communication. The connector connects to a transceiver at each end, so that one fiber communicates in one direction and the other fiber in the other direction.

Two plugs mate in a feed-through receptacle to make a fiber-to-fiber connection. To connect to sources or detectors, we use the connector with a printed circuit board (pcb) receptacle, which mounts on a pcb and houses the active devices (sources and detectors). Active devices having a TO-52 outline fit into bores in the rear of the receptacle. The plug assembly snaps into the front.

Latches on the plug hold it in the receptacle while springs force the termini (the ferrule assemblies on the fiber) into the tapered bores in the receptacle.

FIGURE 11–15 Good and bad polishes (Courtesy of Buehler Ltd., 41 Waukegan Road, Lake Bluff, IL 60044)

AMP ★ OPTIMATE ★ DUPLEX
FIBER-OPTIC CONNECTORS

Instruction Sheet
IS 6856

| RELEASED | 2 · 12 · 85 |

FIBER DIA (μm)	TERMINUS KIT PART NUMBER JACKET DIAMETER (mm) (Max)				PLUG KIT PART NO.
	2.5*†	3.0	3.8	4.4	
125	501165-1	501165-2	501165-3	501165-4	228749-1
140	501166-1	501166-2	501166-3	501166-4	
200	501167-1	501167-1	501167-3	501167-4	
230	501168-1	501168-2	501168-3	501168-4	
250	501169-1	501169-2	501169-3	501169-4	
250**	501170-1	--	--	--	
500**	501171-1	--	--	--	
750**	501172-1	--	--	--	
1000**	501173-1	--	--	--	
Insul Die Set	312831-1	312831-1	312831-2	312831-3	

* FOR PLASTIC FIBERS, JACKET DIAMETER IS 2.2 MM.
** PLASTIC FIBERS.
† REQUIRES SPACER TUBE BE USED WITH TERMINUS.

Fig. 1

1. INTRODUCTION

This instruction sheet (IS) covers the assembly and use of AMP OPTIMATE duplex fiber-optic connectors. The duplex connectors include the following:

— Terminus kits, as listed in Figure 1, which terminate the optical fiber. Two kits must be ordered for each duplex connector.

— Plug Kit 228749-1, which holds two termini.

— PCB Receptacle Kit 228748-1, which houses two TO-52 active devices and mates with the plug.

— Feed-Through Receptacle 228910-1, which accepts two plugs to form a fiber-to-fiber splice.

This sheet is organized as follows:

— Paragraph 2 describes the connector components

— Paragraph 3 provides instructions for applying a terminus kit to an optical fiber.

— Paragraph 4 covers the assembly and application of plugs and receptacles.

— Paragraph 5 contains a note on an alternative method of terminating plastic fibers.

* Trademark of AMP Incorporated

Also available are installation and designer kits to aid in the application procedure. Installation Kits 501258-1 through -4 contain all tools and materials required (see IS 9149 for kit contents). Designer Kit 501219-2 contains an assortment of terminus kits, plugs and receptacles, and tools.

Read this material thoroughly before starting assembly.

2. DESCRIPTION

Figure 1 lists termini kits for various fiber sizes and jacket diameters. Besides the ferrule, into which the fiber is epoxied, each kit contains one each of the following: an eyelet, a compression spring, a polishing bushing, and a dust cap. A spacer tube, used with cables having a jacket diameter of less than .1 in. (2.5 mm), is also included with appropriate kits.

The PCB receptacle consists of the housing and retention clip. The housing accepts active devices having a TO-52 outline. The retention clip secures the housing to the printed-circuit (pc) board. The feed-through receptacle is a single piece that accepts a plug at both ends.

The plug assembly consists of a plug housing, a plug

FIGURE 11-16 Assembly instructions for a duplex connector (Illustration courtesy of AMP Incorporated)

IS 6856

retainer, and a dust cover. The plug retainer features locking latches that secure internal components in the plug housing when mated.

Both plugs and receptacles are polarized so that a plug mates with a receptacle in only one way.

Also required is epoxy. AMP Epoxy 501195-1 has a cure time of 4 hr at 65°C or 24 hr at 25°C. Epoxy 501195-3 has a cure time of 30 minutes at 25°C.

The buffer coating of the fiber can be removed by using the fiber stripper or by dipping the fiber into a heavy-duty paint remover.

The termini are crimped with AMP Hand Tool 58190-1, which must be equipped with the proper insulation crimp die set listed in Figure 1. See IS 9047, packaged with the tool, for further information on the hand crimping tool.

3.   ASSEMBLY PROCEDURE

| WARNING | *Be very careful to dispose of fiber ends properly. The fibers create slivers that can easily puncture the skin and cause irritation.* |

| WARNING | *Always wear safety glasses when working with optical fibers.* |

You can terminate more than one fiber at a time. Do not, however, terminate more than what is practical at a time. Generally, two or three fibers can be terminated before the epoxy begins to gel.

A.   Preparing Fibers

1.   For a dual-channel cable, separate the channels at least 4 in. (10 cm). See Figure 2.

| NOTE | *If the fiber jacket has a diameter over .15 in. (3.8 mm), slip the plug retainer over the jackets before stripping the cable.* |

2.   If the fiber jacket has a diameter of .1 in. (2.5 mm) or less, slip the spacer tube over the jacket.

3.   Slide the compression spring onto the cable.

4.   Slide the eyelet onto the cable.

5.   Strip the cable to the dimensions shown in Figure 2.

| CAUTION | *Do not touch bare fibers with your fingers, because oil from the skin will weaken the fiber by making it brittle.* |

6.   Remove the primary buffer coating from the fiber. This can be done either by soaking the

AMP OPTIMATE DUPLEX CONNECTORS

Fig. 2

end of the fiber in a paint stripper for about 2 minutes or by using the fiber stripper.

7.   Wipe the fiber clean with a soft cloth or tissue.

B.   Terminating Fibers

| NOTE | *When applying epoxy to the ferrule, keep the ferrule vertical with the nose pointing down. This ensures the proper spreading of the epoxy into the front of the nose.* |

1.   Remove the separating clip from the epoxy packet and mix the epoxy thoroughly for 20 to 30 sec. Use of Epoxy Mixer 501202-1 is recommended. Open the tube and pour the epoxy into the foil.

2.   Apply two or three drops of epoxy deep inside the ferrule. See Figure 3 for the points at which to apply the epoxy.

3.   Gather the strength members to one side, and drag the fiber through the epoxy until the fiber is coated.

4.   Insert the fiber into the ferrule with a rotating motion. *Do not force the fiber into the ferrule.* The fiber must extend out the front of the ferrule. The strength members must be outside the ferrule.

FIGURE 11–16   (Continued)

AMP OPTIMATE DUPLEX CONNECTORS   IS 6856

Fig. 3   Fig. 4

5. Press the polishing bushing onto the nose of the ferrule.

6. Gather the strength members to one side, and apply small smears of epoxy to the outside of the ferrule barrel about 1/4 in. (6.3 mm) from the cable entry end.

7. Apply a drop of epoxy to the end of the fiber jacket.

8. Spread the strength members over the ferrule barrel.

9. While holding the ferrule back against the jacket, slide the retaining assembly over the ferrule, trapping the strength members between eyelet and ferrule. See Figure 3.

C. Crimping

1. Squeeze the handles of the hand tool until the ratchet releases. Open the tool fully.

2. Place the terminus into the tool so that the flange of the eyelet rests in the flange of the locator. See Figure 4.

3. Squeeze the handles shut to crimp the eyelet.

4. Set the assembly aside until the epoxy cures. We recommend that the cable be hung, with the terminus down, until the epoxy cures.

D. Polishing the Fiber

Polishing may be done by hand or with AMP Polishing Machine 501186. This sheet describes hand polishing. For information on using the polishing machine, see IS 9107, supplied with the machine.

Fig. 5

WARNING *SAFELY DISPOSE OF THE EXCESS FIBER after it is removed. A handy method is to put a small piece of masking tape on the fiber before scribing it. The fiber is then easily retrieved and disposed of.*

1. With the beveled edge of the scribe tool facing up, lightly scribe the fiber. Then pull the fiber straight away from the polishing bushing. See Figure 5.

NOTE *Plastic fiber can be cut off with a sharp knife. Leave a bit of fiber protruding: Do not cut if flush with the bushing.*

2. Make sure you have a clean surface on

FIGURE 11–16 (Continued)

IS 6856

which to polish the fiber. The polishing surface should be flooded with water, preferably running water.

> CAUTION
> *During polishing, the water and lapping paper must be kept clean to prevent abrasive particles from scratching or chipping the fiber surface.*

3. Place the 5-μm lapping film, dull side up, on the polishing surface.

4. Fine-grind the tip with an elongated figure-8 motion. BEGIN WITH LIGHT PRESSURE, since too much pressure could cause the protruding fiber to fracture or splinter. See Figure 6.

> NOTE
> *Rotate the terminus in your grip occasionally as you polish.*

Fig. 6

AMP OPTIMATE DUPLEX CONNECTORS

5. Fine-grind the tip until the following conditions are met (Figure 7B):
    — The outer pads on the bushing are flush with the flat of the bushing.
    — Polishing may be apparent on the face of the bushing where the high spots have been removed.

6. Replace the 5-μm film with the 1-μm film.

7. Continue polishing on the 1-μm film until the following conditions are met (Figure 7C):
    — The ferrule is flat and has a high-gloss finish.
    — The inner pads are flush with the face of the bushing.
    — Signs of polishing are apparent in a circle around the face of the polishing bushing.

> WARNING
> *Never inspect or look into the end of a fiber when optical power is applied to the fiber. The infrared light used, although it cannot be seen, can injure eyes and cause blindness.*

8. Inspect the ferrule and fiber under the magnifier or microscope. See Figure 8. To view the ferrule properly through the magnifier, light must be reflected off the ferrule and through the magnifier. Check for the following:
    — Deep scratches on the ferrule **indicate** the need for further polishing with 1-μm film. (Fine polishing lines are acceptable.)
    — *Small* chips in the outer rim of the fiber are permissible. Large chips, or chips in the center of the fiber, mean either that further polishing is needed or that the termination is unacceptable and the fiber must be reterminated.

Fig. 7

FIGURE 11–16 (Continued)

AMP OPTIMATE DUPLEX CONNECTORS   **IS 6856**

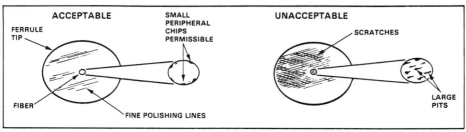

Fig. 8

NOTE   *See IS 9111 for information on the use of AMP Microscope 501196.*

9.   Remove the polishing bushing. Carefully dry the terminus with a soft cloth or tissue.

10.   Place the dust cover over the terminus if the fiber is not going to be installed in the housing immediately.

The terminus is now ready for use.

4.   PLUGS AND RECEPTACLES

A.   Plug Assembly (Figure 9)

1.   Make sure the termini to be used are properly applied and polished.

2.   Insert both termini into the plug housing.

CAUTION   *Ensure that the proper termini are being used and that they are properly located. Make sure each terminus is in the proper side of the retainer at both ends of the cable. Use cable markings or other indicator to ensure this. Once the plug retainer is engaged, it is not removable from the housing.*

3.   With the plug retainer positioned over the fibers as shown in Figure 9, slide the retainer into the plug housing until the locking latches engage the plug housing.

4.   Slide the loaded plug assembly into the receptacle assembly (previously assembled to the pc board) until the latch on the plug assembly engages in the latch hole of the receptacle.

5.   To remove the plug assembly, depress the locking latch through the latch hole and withdraw the plug.

6.   Place the dust cover over unmated assemblies. The cover should NOT cover the latch of the plug assembly.

B.   Receptacle Assembly

Figure 10 shows the recommended layout dimensions for the pc board.

1.   Install the retention clip through the slots provided in the top of the housing.

2.   Install the housing assembly on the pc board by pushing the two plastic posts and the retention clip into the appropriate holes in the pc board. See Figure 11.

3.   Install the active devices in the housing cavities and insert their leads into the appropriate holes in the board.

4.   Hand or wave solder the retention clip posts and the leads of the active device to secure the assembly.

Fig. 9

FIGURE 11-16   (Continued)

IS 6856

Fig. 10

AMP OPTIMATE DUPLEX CONNECTORS

PC BOARD

Fig. 11

C. Feed-Through Receptacle

1. Insert a plug assembly into each plug cavity of the receptacle until the locking latch engages.

2. To remove a plug assembly, depress the locking latch through the latch hole and withdraw the plug.

FIGURE 11–16 (Continued)

SUMMARY

- A splice is used for permanent and semipermanent connections.
- A connector is used for disconnectable connections.
- Loss in an interconnection results from intrinsic factors, extrinsic factors, and system-related factors.
- There are many different designs for connectors and splices. The common ingredient in all designs is precise alignment of fibers.
- Fiber ends can be prepared by the scribe-and-break technique (cleaving) or by polishing.

REVIEW QUESTIONS

1. What is the purpose of a splice or connector?
2. What role does the alignment mechanism play in a fiber-optic connector or splice?
3. Name three sources of intrinsic loss in an interconnection.
4. Name three sources of extrinsic loss in an interconnection.
5. Describe how modal patterns affect loss in a fiber-to-fiber interconnection. Will the loss probably be lower if the transmitting fiber is fully filled or at EMD? Why?

6. Assume an application requiring five 10-km reels of fiber to be joined in a 50-km link. Would connectors or splices be preferred? Why?
7. Would a connector or splice be preferred to connect a fiber-optic cable to a personal computer? Why?
8. Calculate the total mismatch loss for a transmitting fiber with a 62.5/125-μm diameter and a 0.29 NA and a receiving fiber with an 85/125-μm diameter and a 0.26 NA. Ignore other sources of loss for this example.
9. What is the insertion loss test meant to measure? What does the test attempt to eliminate as a loss factor?
10. Name the two critical results a good cleave must achieve.

Chapter 12

Couplers and Networks

Thus far in this book, we have viewed the fiber-optic link as a point-to-point system—one transmitter linked to one receiver over an optical fiber. Even a duplex link is point to point—one transceiver communicates with another transceiver over a fiber-optic pair. In many applications, however, it is desirable or necessary to divide light from one fiber into several fibers or, conversely, to couple light from several fibers into one fiber. In short, we wish to distribute the light. A *coupler* is a device that performs such distribution. Figure 12-1 shows the idea of a coupler: It divides or combines light. The figure also shows examples of actual devices, including couplers, WDM couplers discussed later in this chapter, and transmitters and receivers.

This chapter describes couplers and their uses. Important application areas for couplers are in networks, especially local area networks, and in wavelength-division multiplexing (WDM). To help give perspective to the useful of couplers, we begin the chapter with a description of local area networks. At the end of the chapter, WDM is discussed.

FIGURE 12-1 Directional coupler (Courtesy of Siecor Corporation)

LOCAL AREA NETWORKS

A *local area network* (LAN) is an electronic communication network in which equipment in a limited geographical area, such as an office or building, is interconnected. All equipment is able to communicate with other equipment. In an office, personal computers, word processors, copiers, printers, plotters, facsimile machines, and so forth connect to the network and communicate with one another. It allows workers to share such information as databases, programs, files, and other information, as well as services such as electronic mail.

Since it is estimated that 80% of the information used within a local environment is generated from within that environment, a LAN provides an efficient way to allow users to beneficially share information and resources. LANs are rapidly being installed by businesses, universities, and factories.

A network is composed of a transmission medium that connects several nodes or stations. Each node is a point at which electronic equipment is connected onto the network. A network is a complex arrangement of software and hardware that ensures compatibility not only of signals but also of information.

Most networks were originally designed to use either twisted pairs or coaxial cable as the transmission medium. Most networks have also been adopted to use fiber optics. One network, fiber distributed data interface (FDDI), is expressly designed to use fiber optics.

An important area of a network is its topology. The *topology* defines the physical and logical arrangement. Figure 12–2 shows the most common network topologies.

The bus structure has the transmission medium as a central bus off of which each node is tapped. Messages flow in either direction along the bus.

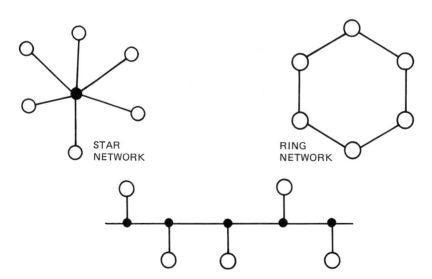

STAR NETWORK

RING NETWORK

FIGURE 12–2 Network topologies

The ring structure has each node connected serially with the one on either side of it. Messages flow from node to node in one direction only around the ring.

The star topology has all nodes interconnected through a central point, through which all messages must pass.

In adapting fiber optics to networks, we sometimes must depart from the physical topology while maintaining the logical topology. This strategy permits fiber optics to be used to full advantage with existing network standards. A fiber-optic-based network may have a *physical* star topology and a *logical* ring topology. This means that the fiber-optic hardware uses a star arrangement, but the equipment attached to the network as well as the software controlling the network sees a ring structure. Couplers are an important consideration in the structure of a network.

COUPLER BASICS

A *coupler* is a multiport device. A *port* is an input or output point for light. There are several types of loss associated with a coupler. Figure 12–3 shows a four-port directional coupler that we will use to define ideas important to your understanding of couplers. Arrows indicate the possible directions of flow for optical power through the coupler. Light injected into port 1 will exit through ports 2 and 3. Ideally, no light will appear at port 4. Similarly, light injected into port 4 will also appear at ports 2 and 3 but not at port 1.

The coupler is passive and bidirectional. Ports 1 and 4 can serve as input ports, ports 2 and 3 as output ports. Reversing the direction of power flow allows ports 2 and 3 to serve as input ports and ports 1 and 4 to serve as output ports.

For the following discussion of loss, we assume port 1 is the *input* port and ports 2 and 3 are the *output* ports. Furthermore, the power at port 2 is always equal to or greater than the power at port 3. Therefore, we will term port 2 as the *throughput* port. Port 3 is the *tap* port. These terms are used to suggest that a path containing the greater part of the power is the throughput path, whereas the path containing the lesser part is the tapped path.

Throughput loss is the ratio of output power at port 2 to input power at port 1:

$$loss_{THP} = 10 \log_{10} \left(\frac{P_2}{P_1} \right)$$

Tap loss is the ratio of the output power at port 3 to the input power at port 1:

$$loss_{TAP} = 10 \log_{10} \left(\frac{P_3}{P_1} \right)$$

Directionality is the ratio between unwanted power at port 4 and the input power at port 1:

$$loss_D = 10 \log_{10} \left(\frac{P_4}{P_1} \right)$$

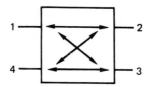

FIGURE 12-3 Four-port directional coupler

Ideally, no power appears at port 4, so $loss_D = 0$. Practically, some power does appear through such mechanisms as leakage or reflections. Directionality is sometimes called *isolation*. Directionality or isolation in a good tap is 40 dB or greater; only a very small amount of light appears at port 4.

Excess loss is the ratio between output power at ports 2 and 3 to the input power at port 1:

$$loss_E = 10 \log_{10} \left(\frac{P_2 + P_3}{P_1} \right)$$

Excess losses are losses that occur because the coupler is not a perfect device. Losses occur within fibers internal to the coupler from scattering, absorption, reflections, misalignment, and poor isolation. In a perfect coupler, the sum of the output power equals the input power ($P_2 + P_3 = P_1$). In a real coupler, the sum of the output power is always something less than the input power because of excess loss ($P_2 + P_3 < P_1$).

Excess loss does not include losses from connectors attaching fibers to the ports. We looked at these losses and their causes in the last chapter. Furthermore, since most couplers contain an optical fiber at each port, additional loss can occur because of diameter and NA mismatches between the coupler port and the attached fiber.

The input power must obviously be divided between the two output ports. The *coupler splitting ratio* is simply the ratio between the throughput port and the tap port: P_2/P_3. Typical ratios are 1:1, 2:1, 3:1, 6:1, and 10:1. From the splitting ratio, the relation of the throughput loss and tap loss is constant. Table 12-1 shows the losses for perfect taps of given splitting ratios.

In a real coupler, the losses at the output ports are the sums of the individual loss and the excess loss. If $Loss_{THP'}$ and $Loss_{TAP'}$ represent throughput and tap loss in a real coupler, actual losses become:

$$Loss_{THP'} = Loss_{THP} + Loss_E$$
$$Loss_{TAP'} = Loss_{TAP} + Loss_E$$

Assume a directional coupler with a 3:1 splitting ratio and an excess loss of 1 dB. Actual losses become 2.25 dB for throughput power and 8 dB for tap power. If 100 μW of optical power are input to port 1, how much output power appears at

SPLITTING RATIO	THROUGHPUT LOSS (dB)	TAP LOSS (dB)
1:1	3	3
2:1	1.8	4.8
3:1	1.25	6
6:1	0.66	8.5
9:1	0.46	10
10:1	0.41	10.4

TABLE 12-1 Losses for ideal four-port directional couplers

each port? Since loss is the ratio of power at the two ports, we know that the throughput power at port 2 equals

$$2.25 \text{ dB} = 10 \log_{10}\left(\frac{P_2}{100 \ \mu W} \right)$$

The equation, notice, accounts for excess loss. The 2.25 dB includes 1.25 dB throughput loss plus 1 dB for excess loss. We rearrange and solve the equation to find the power at port 2:

$$P_2 = \left[\log_{10}^{-1}\left(\frac{-2.25 \text{ dB}}{10} \right) \right] 100 \ \mu W$$

$$= 60 \ \mu W$$

(Notice that since loss is a negative quantity, we make the 2.25 dB negative in the equation.) Tap power at port 3 is

$$P_3 = \left[\log_{10}^{-1}\left(\frac{-7 \text{ dB}}{10} \right) \right] 100 \ \mu W$$

$$= 20 \ \mu W$$

A directional coupler is symmetrical, so that the losses remain the same regardless of which ports serve as the input port, throughput port, tap port, and isolated port.

TEE COUPLER

A *tee coupler* is a three-port device. Figure 12–4 shows the application is a typical bus network. A coupler at each node splits off part of the power from the bus and carries it to a transceiver in the attached equipment. If there are many nodes on the bus, the couplers typically have a large splitting ratio so that only a

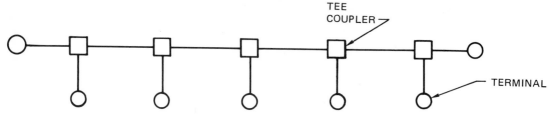

FIGURE 12-4 Tee network

small portion of the light is tapped at each node. The throughput power at each coupler is much greater than the tap power.

Tee couplers are of greatest use when there are a few terminals on the bus. Consider a bus having N terminals. A signal must pass through $N - 1$ couplers before arriving at the receiver. In a coupler having only throughput and tap losses (i.e., no excess loss), the total distribution loss is

$$L = (N - 1)\text{loss}_{THP} + \text{loss}_{TAP}$$

Loss increases linearly with the number of terminals on the bus.

Unfortunately, we must also account for excess loss and connector loss L_C (including any loss from diameter and NA mismatches) at each coupler. Since a connector is required at both the input and output ports of the coupler, the number of required connectors is $2N$. These losses also add linearly with the number of terminals, so the real total distribution loss becomes

$$L = (N - 1)\text{loss}_{THP} + \text{loss}_{TAP} + 2NL_C + \text{loss}_E$$

As the number of terminals added to a network using tee couplers increases, losses mount quickly. As a result, tee couplers are only useful when a small number of terminals are involved. The difference in losses between an ideal network (having only throughput and tap losses) and a real network (having also excess and connector loss) rapidly becomes greater.

The network in Figure 12-4 is one directional. A transmitter at one end of the bus communicates with a receiver at the other end. Each terminal also contains a receiver. A duplex network can be obtained by adding a second fiber bus. It can also be obtained by using an additional directional coupler at each end and at each terminal. Such additions allow signals to flow in both directions. Figure 12-5 shows an example.

The loss characteristics of a tee network require receivers to have large dynamic ranges. Consider a 10-terminal network with terminal 1 serving as the transmitter. The power received at terminal 2 and at terminal 10 will be quite significantly different. Thus, the dynamic range of the receiver in each terminal must be large. If we assume a coupler with a 10:1 splitting ratio, assume excess and connector losses at each coupler total 2 dB, and ignore all other losses in the system, the difference between received power at terminal 1 and at terminal 10 is about 30 dB. The receivers require a dynamic range of 30 dB or better. Other real losses will, of course, increase the requirement.

Notice that failure of a single coupler does not shut down the entire network. It simply divides the network into two smaller networks, one on each side of the failed coupler.

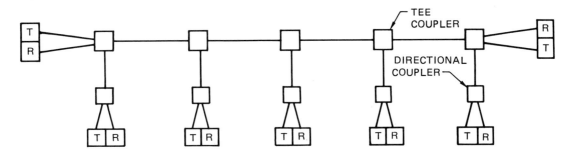

FIGURE 12-5   Tee network with directional coupler

STAR COUPLER

The *star coupler* is an alternative to the tee coupler that makes up for many of the latter's drawbacks. Figure 12-6 shows a transmissive star coupler that has an equal number of input and output ports. For a network of N terminals, the star coupler has $2N$ ports. Light into any input port is equally divided among all the outport ports.

The insertion loss of a star coupler is the ratio of the power appearing at a given output port to that of an input port. Thus the insertion loss varies inversely with the number of terminals:

$$loss_{IN} = 10 \log_{10} \frac{1}{N}$$

Loss does not increase linearly with the number of terminals.

If we add excess loss and connector loss, the total distribution loss becomes

$$L = 10 \log_{10} \frac{1}{N} + loss_E + 2L_C$$

As a result, star couplers are more useful for connecting a large number of terminals to a network.

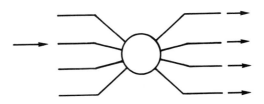

FIGURE 12-6   Star coupler

Ideally, the light is coupled evenly into all the output ports. Practically, it is not. The actual amount of power into each output port varies somewhat from the ideal determined from the insertion loss. *Uniformity* is the term used to specify the variation in output power at an output port. Uniformity is expressed either as a percentage or in decibels. Consider a coupler in which the output power at each port is $50\mu W$. A uniformity of ±0.5 dB means that the actual power will be between 45 and 56 μW. If the uniformity figure increases to ±1 dB, the output varies between 40 and 63 μW.

Figure 12-7 compares loss versus the number of terminals for ideal and real tee and star couplers. The figure assumes an excess loss of 1 dB and a connector loss of 1 dB for each connector, for a total of 3 dB. The figure illustrates the important difference between tee and star couplers. Excess and connector losses occur at each coupler. Since a star network requires only one coupler, these losses only occur once. With a tee network, they increase linearly with the number of couplers in the system. Loss in a star network increases much slower than in a tee network as the number of terminal increases.

One reason for using a tee-coupled network instead of a star-coupled network is that the tee network requires less cable. The centralized location of a star coupler requires significantly more cable to connect widely separated terminals.

Figure 12-8 shows a block diagram of a typical star network using a 4 × 4 star coupler.

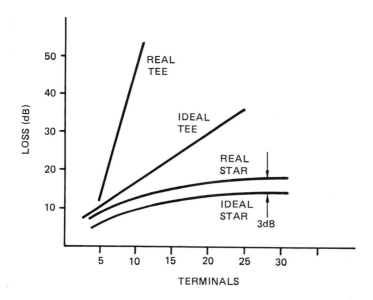

FIGURE 12-7 Loss versus number of terminals for tee and star couplers

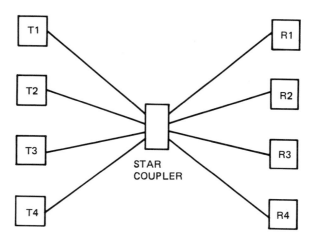

FIGURE 12-8   Star coupler application

REFLECTIVE STAR COUPLERS

A *reflective star coupler* contains N number of ports in which each port can serve as an input or an output. Light injected into any one port will appear at all other ports.

COUPLER MECHANISMS

This section looks at several means of constructing a coupler. Most couplers are "black boxes," in which the internal coupling mechanisms are of little interest to the user. Typical input/output ports to the coupler are either fiber pigtails or connector bushings. The advantage of the pigtail is that it permits greater flexibility of application, since connection to the pigtails can be made with any compatible splice or connector. Couplers using connectorized ports are naturally limited to interconnection by specific connector type. The connectorized couplers are, on the other hand, ready to go; inexperienced users can simply connect or disconnect the fiber to the coupler.

Fused Couplers

A *fused star coupler* is made by wrapping fibers together at a central point and heating the point. The glass will melt into a unified mass so that light from any single fiber passing through the fused point will enter all the fibers on the other side. A transmissive star coupler results when an end of each fiber is on each side of the fused section. A reflective star results when the fibers loop back so that each fiber is twice fused in the fused section. Both types are shown in Figure 12-9.

TRANSMISSIVE STAR

REFLECTIVE STAR

FIGURE 12-9 Fused star couplers

Mixing Rods

A *mixing rod* is a relatively large, rectangular or circular waveguide. The rectangular mixer resembles an optical fiber in that its top and bottom typically have a layer of material with a lower refractive index. A short length of large-core fiber can also serve as a mixer. Light from the small core of an attached fiber mixes throughout the mixer.

In a transmissive star coupler, input fibers butt against one end of the rod and output fibers against the other. Light into the mixing rod is mixed throughout the rod so that it enters all the fibers at the output end.

In a reflective star coupler, all fibers butt against one end of the mixing rod. The other end is coated with a highly reflective material. Light from one fiber into the rod reflects back into all the fibers. Both types of star couplers are shown in Figure 12-10.

Optical Approach

Optical approaches to couplers include the beamsplitter shown in Figure 12-11. The *beamsplitter* is a plate coated with a partially reflective material. A portion of the light transmits through the plate and a portion is reflected. The ratio of transmitted and reflective light is controlled by the coating. The beamsplitter by itself is separated by a large gap from the fibers, which causes unacceptable

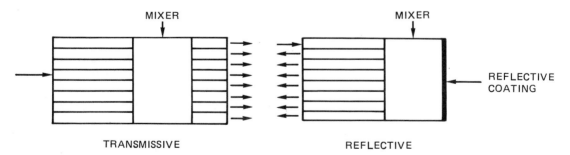

MIXER

MIXER

REFLECTIVE COATING

TRANSMISSIVE

REFLECTIVE

FIGURE 12-10 Star couplers with mixing rod

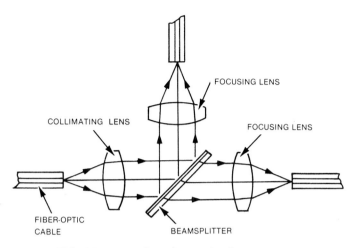

FIGURE 12–11   Coupling with a beamsplitter

losses. Collimating and focusing lenses control the path of the light and reduce losses internal to the coupler.

Discrete Fibers

A coupler can also be constructed of discrete fibers. Figure 12–12 shows an example of a seven-way splitter/combiner that uses the simplex connector discussed in the last chapter. The connector at the bundled end has a ferrule that holds seven 100/140-μm fibers. The other end of each fiber is terminated with a separate connector. The diameter of the cores of the bundled end is 380-μm, allowing it to be connected to a large-core fiber, to a source with a large emitting diameter, or to a detector with a large receiving area. Losses are those associated with each individual fiber in the bundle. Calculations are based on the connected element (source, detector, or fiber) and each individual fiber. For example, in coupling light from a source into the bundled end, losses are calculated for NA and diameter mismatches between the source and a 100/140 fiber. A source with a 400-μm-diameter emitting area results in a loss of 2 dB from diameter mismatch:

$$\text{loss}_{\text{dia}} = 10 \log_{10} \left(\frac{100}{400} \, \mu\text{m} \right)^2$$

$$= -12 \text{ dB}$$

These losses are comparable with those of other coupler constructions.

As shown in the figure, this approach can be used for either squid or unitized couplers. *Squid* types use individual, free cables to form the coupler. *Unitized* versions are small packages. Different sizes or numbers of fibers can be accommodated by changing the size of the ferrule hole in the bundled end.

380-μm
ACTIVE AREA

420-μm-DIA
BUNDLE

100 μm CORE

20 μm CLADDING

140-μm
DIA
FIBER

Type	Source	Fiber Core Dia	Fiber Cladding Dia	Bundled End Core Dia
	nm	μm	μm	μm
1x2 Squid	820	85	125	210
1x3 Squid	820	85	125	229
1x7 Squid	820	100	140	380
1x7 Unitized	820	100	140	380
1x7 Unitized	660	240	250	730
1x7x1 Star Unitized	660	240	250	730

FIGURE 12–12 Splitter/combiner using discrete fibers (Illustration courtesy of AMP Incorporated)

WAVELENGTH-DIVISION MULTIPLEXER

Multiplexing is a method of sending several signals over a line simultaneously. In Chapter 2, we saw how telephone companies use time-division multiplexing to send hundreds of telephone calls over a single line. Each voice is first transformed in digital data by pulse-coded modulation (PCM). Each PCM-encoded signal is then allotted specific time slots within the transmission. Actual modulation and demodulation are accomplished electrically, before being presented to the fiber-optic transmitter and after being received from the fiber-optic receiver.

Wavelength-division multiplexing (WDM) uses different wavelengths to multiplex two or more signals. Transmitters operating at different wavelengths can each inject their optical signals into an optical fiber. At the other end of the link, the signals can again be discriminated and separated by wavelength. A WDM coupler serves to combine separate wavelengths onto a single fiber or to split combined wavelengths back into their component signals. The record 20-Gbps fiber-optic transmission mentioned in Chapter 3 used WDM to combine the outputs of 10 lasers, each operating at 2 Gbps, into a single-mode fiber. The lasers had output wavelengths very close to one another in the 1550-nm region.

Figure 12–13 shows an example of WDM. Two transmitters, one operating at 820 nm and one at 1300 nm, both present signals to a WDM coupler, which couples both signals onto a fiber. At the other end, a second WDM coupler separates the received light back into its 820- and 1300-nm components and presents it to two receivers.

Two important considerations in a WDM device are crosstalk and channel separation. Both are of concern mainly in the receiving or demultiplexing end of the system. *Crosstalk* refers to how well the demultiplexed channels are separated. Each channel should appear only at its intended port and not at any other output port. The crosstalk specification expresses how well a coupler maintains this port-to-port separation. Crosstalk, for example, measures how much of the 820-nm wavelength appears at the 1300-nm port. A crosstalk of 20 dB means that 1% of the signal appears at the unintended port.

Channel separation describes how well a coupler can distinguish wavelengths. In most couplers, the wavelengths must be widely separated, such as 820 nm and 1300 nm. Such a device will not distinguish between 1290-nm and 1310-nm signals.

Most WDM devices today operate with only a few separate optical channels. Practical, affordable couplers do not discriminate between closely separated channels. A three-port coupler, for example, might multiplex 820-nm and 1300-nm wavelengths, whereas a four-port coupler multiplexes 755-, 820-, and 1300-nm wavelengths.

The limited number of channels for WDM from the spectral requirements of a multichannel WDM device. LED sources have wide spectral widths, which require each channel to be separated widely from the others. Constructing a

FIGURE 12–13   Wavelength-division multiplexing

WDM device with many channels requires a narrow band source. Lasers, which have sufficiently narrow spectral widths, are not readily available within the range of outputs required for a high number of channels. Single-frequency lasers with a tunable wavelength have been demonstrated in the laboratory, but they are not yet commercially available.

WDM allows the potential information-carrying capacity of an optical fiber to be increased significantly. The bandwidth-length product used to specify the information-carrying capacity of a fiber applies only to a single channel—in other words, to a signal imposed on a single optical carrier. A fiber with a 500-MHz-km bandwidth used with a four-port WDM coupler can carry a 500-MHz signal on the 755-nm channel, a 500-MHz signal on the 820-nm channel, and a 500-MHz signal on the 1300-nm channel. Its effective information-carrying capacity has been increased to 1500 MHz.

The two approaches to WDM are diffraction gratings and interference filters. Although there are many ways to achieve each in practical application, Figure 12–14 shows a single example of each. The *diffraction grating* is an array of fine, parallel reflecting lines spaced on the order of a wavelength. Because of the small size of the grating, the laws of reflection and refraction are replaced by those of diffraction. *Diffraction* is caused by the interaction of the wave and an object, in this case the grating. Diffraction causes deviation of waves from their paths. The grating concentrates the diffracted light to a few directions. The number of parallel lines and their spacings determine the number of component channels. The grating also works in the reverse direction to combine wavelengths. The diffraction grating is most useful for systems with 5 to 10 channels. Since a correspondingly broad spectrum of sources is not available, diffraction-grating WDM is not widely available either.

An optical filter can also be used to separate wavelengths. A *dichroic* substance is one that reflects or transmits light selectively according to wavelength. The multiplexer shown in Figure 12–14 has three optical fibers arranged precisely as shown. One of the in-line fibers is coated with a dichroic substance to control the direction of light on a wavelength-sensitive basis. Although most such multiplexers will use wavelengths such as 820 nm and 1300 nm, modification of the dichroic coating permits separation of wavelengths as closely as 100 nm.

FIGURE 12–14   WDM mechanisms

ACTIVE COUPLER

The couplers we have discussed so far are passive. A passive device is unpowered, usually causes loss, and usually works in either direction. In electronics, passive devices include resistors, capacitors, and inductors. An active device, on the other hand, usually depends on some application of power to cause a change in its state; it is also usually sensitive to direction of applied power. An LED works differently, depending on whether the bias is forward or reverse. Examples of active electronic components include diodes, transistors, LEDs, and photodetectors.

Figure 12–15 shows an example of an active coupler. It consists internally of a source, detector, beam divider, focusing mirror, and pigtailed fiber. Light from

FIGURE 12-15 Active coupler for duplex transmission over a single fiber (Courtesy of ADC Fiber Optics Corporation)

the source is coupled by the beam divider into the fiber. Light from the fiber passes through the beam divider and is directed by the focusing mirror onto the detector.

Incorporating both a source and detector along with the coupling mechanisms in a single package allows transmission in both directions along a single fiber. Operation is full duplex, since both the source and detector can operate simultaneously. Electrical leads for the optoelectronic device permit mounting on a printed circuit board.

OPTICAL SWITCH

It is desirable at times to couple light from one fiber to one of two fibers but not to both. A passive coupler does not allow such a choice: The division of light is always the same. An optical switch does allow such choice. It is analogous to an electrical switch, since it permits one of two circuit paths to be chosen, depending on the switch setting. Figure 12–16 shows a possible construction of an electrically operated switch as well as a typical application as a bypass switch.

One network configuration is a ring in which terminals are daisy-chained. In other words, the transmitter of the first terminal connects to the receiver of the second terminal. The second terminal's receiver, in turn, feeds the transmitter,

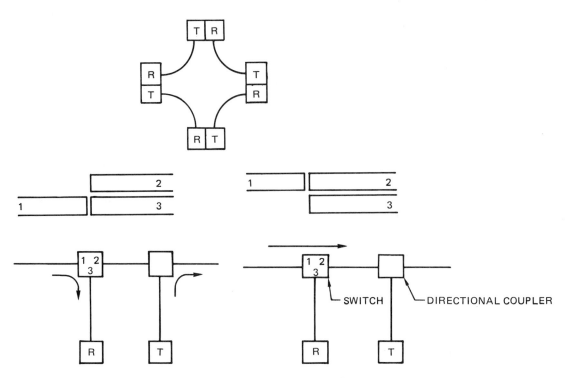

FIGURE 12–16 Fiber-optic switch

which connects to the receiver of the third terminal. This arrangement continues from terminal to terminal around the ring.

The network is a series of point-to-point links. It has the advantage of a simple arrangement. Losses within the system are limited to those of each point-to-point section. The complexities of loss associated with a tee-coupler based bus network are avoided. The disadvantages of this ring arrangement is that failure of a single terminal shuts down the entire network. We cannot even turn off a terminal without disrupting the network.

The fiber-optic bypass switch overcomes this problem. The two settings of the switch permit the light signal to be transmitted to the terminal receiver or to bypass the terminal and continue on the ring to the next terminal. A directional coupler after the switch must also be used in conjunction with the switch.

The switch uses a relay arrangement to move the fiber between positions. A switch can be constructed so that it automatically switches to the bypass position if power is removed, either from turning off the terminal intentionally or from unexpected disruption. The result is a certain degree of "fail-safe" operation.

The drawback to such switches is the difficulty of manufacturing low-loss switches. We have already discussed the tight tolerances imposed by the need for precise alignment in fiber optics. Maintaining alignment on moving parts and over repeated switchings simply compounds the difficulty.

SUMMARY

- A coupler is used to distribute light.
- The two principal types of couplers are tee couplers and star couplers.
- A network is a multipoint fiber-optic system.
- A tee coupler is most useful in a network containing few terminals.
- A star coupler is most useful in a network containing many terminals.
- Wavelength-division multiplexing is a method of multiplexing two or more optical channels separated by wavelength.
- A fiber-optic switch permits light from one fiber to be switched between two or more fibers.

REVIEW QUESTIONS

1. Name the two main types of couplers.
2. Sketch the operation of a three-port directional coupler.
3. Explain why a tee coupler is more useful when used in a network with few terminals. Give two advantages of a tee coupler over a star coupler.
4. List three network topologies. Sketch a six-terminal example of each.
5. Calculate the loss for a 4 × 4 star coupler connected to four terminals, and for an 8 × 8 star coupler connected to eight terminals. Ignore excess and connector losses.

6. What is the difference between transmissive and reflective star couplers?
7. Assume a fiber has a 600-MHz-km bandwidth and, through WDM, transmits optical channels at 850 nm and 1300 nm. For a 1-km link, what is the highest bandwidth of each channel? Why?
8. What is the total information-carrying capacity of the fiber in Question 7?
9. Describe how the application of a fiber-optic switch differs from the operation of a passive coupler.

Part Three

FIBER-OPTIC SYSTEMS

Chapter 13

The Fiber-Optic Link

We have now looked at the main components of a fiber-optic link, at cables, sources and transmitters, detectors and receivers, and connectors and couplers. This chapter looks at how these components are brought together into a link and at how a link is planned. It details the ingredients of a power budget and a rise-time budget, which are the two fundamental requirements that ensure that a link meets its intended application requirements. We will learn how to construct and evaluate such budgets for a simple fiber-optic system.

PRELIMINARY CONSIDERATIONS

The first step in any application is to have an application suitable to fiber optics in the first place. Wiring your doorbell with a fiber-optic link might be an interesting exercise, but it is hardly practical. Similarly, linking a personal computer to a nearby printer is most easily and inexpensively done with a standard electrical cable. If, however, the printer is a great distance away or there are strong noise sources nearby that interfere with transmission, fiber optics becomes of interest.

The decision to use fiber optics involves comparing its advantages, disadvantages, and costs against competing copper solutions. We will, however, assume fiber optics has been chosen as the best medium.

The next step is to decide whether to buy a complete fiber-optic system, to use packaged transmitters and receivers and build your own system from there, or to design and build your own transmitters and receivers as well. Because of our interests in this chapter, we select the middle choice. The third option sidetracks us from the main purpose of this chapter, which is budgeting the optical portion of the link. The first option, which will be installed by a vendor or contractor, leaves us with nothing to discuss.

SYSTEM SPECIFICATIONS

In planning a fiber-optic system, we must define our application requirements so that we can specify our needs. The main question involves the data rates and distances involved: how far? how fast? These questions give us the basic application constraints. Beyond that, we must specify the BER required.

Now that we have the main requirements—distance and data rate—we can begin to evaluate the other factors involved:

- Type of fiber
- Operating wavelength
- Transmitter power
- Source type: LED or laser
- Receiver sensitivity
- Detector type: pin diode, APD, IDP
- Modulation code
- BER
- Interface compatibility
- Number of connectors
- Number of splices
- Environmental concerns
- Mechanical concerns
- Other special concerns

The environmental and mechanical concerns involve such issues as temperature and humidity ranges, indoor/outdoor application, flammability requirements, and so forth. These factors will especially affect the choice of fiber-optic cable.

We can see that many of these questions are related and cannot be as easily separated as we have done here. The receiver sensitivity is influenced by the choice of detector. The receiver sensitivity sets the minimum optical power required by the receiver. The power arriving at the receiver, however, depends on the transmitter power and the fiber attenuation. We can specify a very sensitive receiver that will allow us to use a transmitter of lower power. Or we can use a less sensitive receiver, in which case our transmitter must be more powerful.

Consider a given transmitter, fiber-optic cable, and receiver. Now assume the power at the receiver is insufficient to meet the BER requirement of the application. We actually have five choices to remedy the situation:

1. A transmitter with a higher output power
2. A fiber with lower attenuation
3. A shorter transmission distance, to lower losses along the fiber length
4. A receiver with lower sensitivity level
5. A modification of the system requirements, so that a lower BER is acceptable (a lower BER implies a lower minimum power at the receiver)

The point is that planning a fiber-optic link is not a cut-and-dried, step-by-step procedure. However, there are logical and rational ways to proceed. One approach is the link power budget.

POWER BUDGET

The *power budget* provides a convenient way to analyze and quantify losses in a link. The basic task of the link is to deliver enough of the transmitter's power to the receiver. The budget is the difference between the transmitter power and the receiver sensitivity. If the peak transmitter power is −10 dBm and the receiver sensitivity is −30 dBm, the power budget is 20 dB. The link cannot tolerate more than 20 dB of loss if it is to operate properly. We must then analyze all losses within the link to ensure that their total does not exceed 20 dB. If the total is less than 20 dB, the remaining power is our margin. If, for example, the total loss is 14 dB, the margin is 6 dB.

A prudently planned link has a power margin. The power margin reserves power for declines in transmitter power due to aging of the source, additional losses from link repairs (as by splicing a broken fiber), and losses that might occur from cable bends or mechanical stresses. Typical margins are 3 to 6 dB.

Figure 13–1 is a graphical representation of a power budget. The graph depicts the power levels at different points in the link to finally show the power arriving at the receiver. The difference between that power and the receiver sensitivity is the margin.

Let us assume the following numerical values for the link in Figure 13–1:

Transmitter power	−10 dBm (100 μW)
Fiber attenuation	3 dB/km
Link length	2 km

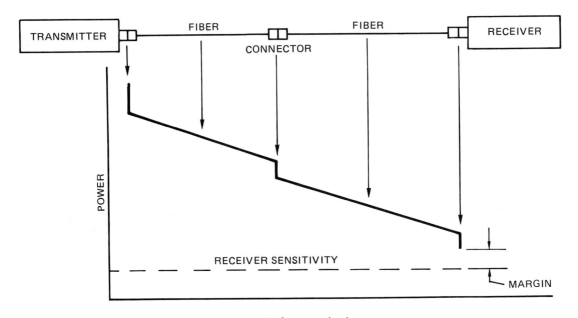

FIGURE 13-1 Link power budget

Connector loss   1.5 dB each interconnection
Receiver sensitivity   –30 dBm (1 μW)

The total loss from fiber attenuation is 6 dB for the 2-km run. Total loss between transmitter and receiver is 7.5 dB, leaving a margin of 12.5 dB.

A MORE COMPLEX EXAMPLE

The link power budget just shown was very straightforward. Let us take a look at a more complex example. The basic layout is the same as the link in Figure 13–1, but now we must calculate the power coupled from the transmitter into the fiber and from the fiber into the receiver. In addition, we will assume the lengths of fiber are different types of fiber. We will require a 6-dB power margin. The question is, how long can the second length of fiber be?

Here are the pertinent specifications:

Transmitter
Output power   250 μW (–6 dBm)
Output diameter   100 μm
NA   0.30
Connector loss   1 dB

Fiber 1
Size   85/125 μm
NA   0.26
Attenuation   6 dB/km
Length   2 km

Connector
Loss (max)   1.5 dB

Fiber 2
Size   100/140 μm
NA   .3
Attenuation   5 dB/km
Length   ?

Receiver
Sensitivity   125 nA (–39 dBm)
Diameter   150 μm
NA   0.4
Connector loss   1 dB

A brief inspection of the specifications shows many instances of the diameter and NA mismatches discussed in this book. The example is a practical one, however. The specifications are typical of those encountered in assembling real components. They were not chosen simply to complicate the example with mismatches: The mismatches occur easily enough in the real world. The mismatches were not included in the first example because the transmitter and receiver manufacturer already accounted for those when the device was specified.

The link power budget is 33 dB, since the difference between the –39-dBm receiver sensitivity and the –6-dBm transmitter power is 33 dB. After subtracting the 6-dB margin, we have 27 dB remaining for losses in the link. We then calculate all other losses in the link. The remaining loss is then available for fiber 2.

Transmitter Losses

Transmitter losses are from NA mismatch, diameter mismatch, and insertion loss from the connector, which is 1 dB. Loss from diameter mismatch between the 100-μm-diameter transmitter output and the 85-μm-diameter fiber is

$$\text{loss}_{dia} = 10 \log_{10} \left(\frac{\text{dia}_{fiber}}{\text{dia}_{tr}} \right)^2$$

$$= 10 \log_{10} \left(\frac{85}{100} \right)^2$$

$$= -1.4 \text{ dB}$$

Loss from NA mismatch between the 0.30 NA of the transmitter and the 0.26 of the fiber is

$$\text{loss}_{NA} = 10 \log_{10} \left(\frac{\text{NA}_{fiber}}{\text{NA}_{tr}} \right)^2$$

$$= 10 \log_{10} \left(\frac{0.26}{0.30} \right)^2$$

$$= -1.2 \text{ dB}$$

The loss at the transmitter interface is 3.6 dB. The total loss is now 3.4 dB.

Fiber 1 Loss

Since the attenuation for the 85/125 fiber is 5 dB/km, the loss for a 2-km run is 10 dB. The total loss is now 13.6 dB.

Fiber-to-Fiber Connection

Since fiber 2 has a larger core diameter and NA than fiber 1, no mismatch losses occur. The only loss is connector insertion loss of 1.5 dB. The total loss now is 15.1 dB.

Receiver Losses

The loss at the receiver is simply the 1 dB from the connector. Since the diameter and the NA of the detector in the receiver is larger than the diameter and NA of the fiber, no mismatch loss occurs. The total loss is now 16.1 dB.

FIGURE 13–2 Link power budget

Fiber 2 Loss

The losses calculated so far total 16.1 dB. Since allowed losses were 27 dB, we have 10.9 dB left for attenuation in the 100/140 fiber. This fiber has an attenuation of 5 dB/km, so we can run it 2.2 km. Figure 13–2 shows this application, and the losses and power levels at each point in the system.

If, in this example, the length of fiber 2 was set at 0.5 km, we would have 8.4 dB left in our budget. We could reduce the power output of the transmitter to extend the life of the source. Or we could operate the receiver above the required sensitivity to obtain a better BER.

The budget can be written in tabular form as shown in Figure 13–3.

Transmitter power	–6 dBm	
Receiver sensitivity	–39 dBm	
Power budget		**33 dB**
Transmitter losses	3.6 dB	
Fiber 1 loss (2 km)	10 dB	
Fiber-to-fiber loss	1.5 dB	
Fiber 2 loss (2.2 km)	10.9 dB	
Receiver loss	1 dB	
Total loss		27 dB
Power margin		6 dB

FIGURE 13–3 Power budget in tabular form

ADDED COMPLEXITIES

A power budget is complicated by the facts of the real world. Many components have a range within their specifications. The peak output of a transmitter may be specified with a 3-dB difference between its minimum and its maximum. The connector loss used was a 1.5-dB maximum. A typical value might be closer to 0.6 dB, and there is the possibility it might reach 0.2 dB. Fiber attenuation varies somewhat with temperature, usually decreasing from that specified. Therefore, a link budget should be calculated for minimum and maximum values to obtain worst-case and best-case values.

In some cases, a link that operates satisfactorily under worst-case conditions of minimum transmitter peak power and maximum losses could overdrive the receiver under best-case conditions of maximum transmitter peak power and minimum losses. Such problems most likely will occur if the receiver has a small dynamic range. If the receiver becomes overdriven, the drive current to the transmitter can be reduced, or additional attenuation can be introduced into the system. Connectors used in the telephone industry, for example, have barrels that fit over them. The sole purpose of the barrel is to provide additional attenuation. These barrels are added as needed.

RISE-TIME BUDGET

The power budget analysis ensures that sufficient power is available throughout the link to meet application demands. Power is one part of the link requirement. The other part is bandwidth or rise time. All components in the link must operate fast enough to meet the bandwidth requirements of the application. The rise-time budget allows such analysis to be performed.

As we saw in the chapters on sources and detectors, active devices have finite response times in response to inputs. The sources and detectors do not turn on and off instantaneously. Rise and fall times determine response time and the resulting bandwidth or operating speed of the devices.

Similarly, dispersion limits the bandwidth in the fiber. In a single-mode fiber, only material dispersion from the source spectral width severely limits bandwidth. In a multimode fiber, the delay caused by light traveling in different modes also limits bandwidth or cable rise time.

When the bandwidth of a component is specified, its rise time can be approximated from

$$t_r = \frac{0.35}{BW}$$

This equation accounts for modal dispersion in a fiber. The rise time must be scaled to the fiber length of the application. If the cable is specified at 600 MHz/km, and the application is 2 km, the bandwidth is 300 MHz and the rise time is 1.6 ns.

In addition, the rise-time budget must include the rise times of the transmitter and receiver, which are typically specified for packaged devices. Transmitter and receiver rise times are used instead of source and detector rise times, since the transmitter and receiver circuits will limit the maximum speed at which optoelectronic devices can operate. For the receiver, rise time/bandwidth may be limited by either the rise time of the components or by the bandwidths of the *RC* time constant.

Connectors, splices, and couplers usually do not affect system speed, and they do not have to be accounted for in the rise-time budget.

When all individual rise times have been found, the system rise time is calculated from the following:

$$t_{rsys} = 1.1 \sqrt{t_{r1}^2 + t_{r2}^2 + \cdots + t_{rn}^2}$$

The 1.1 allows for a 10% degradation factor in the system rise time. In a typical application, the rise-time budget can also be used to set the rise time of any individual components, such as the transmitter or cable, by rearranging the equation to solve for the unknown rise time.

When rearranging the equation to solve for an individual rise time, the 1.1 factor is not used. Since, after all, the factor allows for degradation within the entire system, it should not be applied to individual components.

Consider as an example a 20-MHz application operating over 2 km. The fiber used has a 400-MHz-km bandwidth. The receiver rise time is 10 ns. What is the rise time required of the transmitter?

The required system rise time is 17.5 ns. Fiber rise time is 1.75 ns. Solving for the transmitter rise time gives

$$t_{rtrans} = \sqrt{17.5^2 - 10^2 - 1.75^2}$$
$$= 14.25 \text{ ns}$$

The transmitter must have a rise time of about 14 ns. If we select a transmitter with a rise time of 10 ns, we will meet the rise time or bandwidth requirements of the application. With a 10-ns transmitter, the system rise time becomes

$$t_{rsys} = 1.1 \sqrt{10^2 + 1.75^2 + 10^2}$$
$$= 15.7 \text{ ns}$$

which is within the required 17.5 ns.

SUMMARY

- A power budget ensures that losses are low enough in a link to deliver the required power to the receiver.
- A rise-time budget ensures that all components meet the bandwidth/rise-time requirements of the link.

REVIEW QUESTIONS

1. Using the transmitter and receiver specifications in Table 10–1 and performing a power budget analysis, determine the longest length of fiber that can be used. The fiber used is a 50/125 fiber with a bandwidth of 600 MHz and loss of 4 dB/km. Assume a 6-dB margin. Assume that a fiber-to-fiber connection with a 1-dB loss will be required after every 2 km. Use the maximum output value for the transmitter and the minimum value for the receiver.

2. Sketch a power budget graph for Question 1.

3. Repeat the analysis from Question 1 for a 100/140 fiber with a bandwidth of 400 MHz and loss of 6 dB/km.

4. Sketch a power budget for Question 3.

5. Determine the system rise time for the link analyzed in Question 1. Use the length of fiber determined from the power budget. Use the maximum rise time values for the transmitter and receiver.

Chapter 14

Fiber-Optic Cable Installation and Hardware

This chapter looks at some of the factors involved in installing a fiber-optic cable system. These factors are of interest because they not only demonstrate the practical importance of the mechanical properties of a cable, but they also demonstrate the practical aspects of dealing with fiber optics. We will also describe some types of common fiber-optic hardware, such as closure/organizers, rack boxes, and distribution panels. This hardware is an important part of more complex fiber-optic systems.

BEND RADIUS AND TENSILE RATING

Because of their light weight and extreme flexibility, fiber-optic cables are often more easily installed than their copper counterparts. They can be more easily handled and pulled over greater distances.

Minimum bend radius and *maximum tensile rating* are the critical specifications for any fiber-optic cable installation, both during the installing process and during the installed life of the cable. Careful planning of the installation layout will ensure that the specifications are not exceeded. In addition, the installation process itself must be carefully planned. This chapter discusses the factors involved in the planning.

The minimum bend radius and maximum tensile loading allowed on a cable differ during and after installation. An increasing tensile load causes a reversible attenuation increase, an irreversible attenuation increase, and, finally, cracking of the fiber. The tensile loading allowed during installation is higher than that allowed after installation. Care must be taken in either case not to exceed the specified limits.

The minimum bend radius allowed during installation is larger than the bend radius allowed after installation. One reason is that the allowed bend radius increases with tensile loading. Since the fiber is under load during installation, the minimum bend radius must be larger. The allowed bend radius after installation depends on the tensile load.

Figure 14–1 shows the cross section of a simplex fiber and a duplex fiber, which we will use as examples in this chapter, and gives the specifications for their minimum bend radius and maximum tensile rating.

As discussed in Chapter 6, outdoor cables are commonly multifiber cables with constructions more complex than indoor cables. Indoor cables are typically simplex or duplex.

DIRECT BURIAL INSTALLATION

Cables can be buried directly in the ground by either plowing or trenching methods. The plowing method uses a cable-laying plow, which opens the ground, lays the cable, and covers the cable in a single operation. In the trench method, a trench is dug with a machine such as a backhoe, the cable is laid, and the trench is filled. The trench method is more suited to short-distance installations.

Buried cables must be protected against frost, water seepage, attack by burrowing and gnawing animals, and mechanical stresses that could result from earth movements. Armored cables specially designed for burial are available. Cables should be buried at least 30 in. deep so they are below the frost line. Other buried cables should be enclosed in sturdy polyurethane or PVC pipes. The pipes should have an inside diameter several times the outside diameter of the cable to protect against earth movements. An excess length of cable in the pipe prevents tensile loads being placed on the cable.

AERIAL INSTALLATION

Aerial installation includes stringing cables between telephone poles or along power lines. Unlike copper cables, fiber-optic cables run along power lines with no danger of inductive interference.

Aerial cables must be able to withstand the forces of high winds, storms, ice loading, and so forth. Self-supporting aerial cables can be strung directly from pole to pole. Other cables must be lashed to a high-strength steel wire, which provides the necessary support. The use of a separate support structure is the usual preferred method.

INDOOR INSTALLATION

Most indoor cables must be placed in conduit or trays. Since standard fiber-optic cables are electrically nonconductive, they may be placed in the same ducts as high-voltage cables without the special insulation required by copper wire. Many cables cannot, however, be placed inside air conditioning or ventilation ducts for the same reason that PVC-insulated wire should not be placed in

SIMPLEX CABLE CROSS SECTION

DUPLEX CABLE CROSS SECTION

TENSILE LOADING (MAX)	CABLE
DURING AFTER	90 lb (400 N) 11.25 lb (50 N)
BEND RADIUS (MIN)	CABLE
DURING	5.9 in (150 mm)
AFTER (NO LOAD) AFTER (FULL LOAD)	1.2 in (30 mm) 4.0 in (130 mm)

FIGURE 14-1 Bend radius and tensile strength loading for cable examples (Courtesy of Canoga Data Systems)

these areas: A fire inside these ducts could cause the outer jacket to burn and produce toxic gases.

Plenum cables, however, can be placed in any plenum area within a building without special restrictions. The material used in these cables does not produce toxic fumes.

TRAY AND DUCT INSTALLATIONS

The first mechanical property of the cable that must be considered in planning an installation is the outside diameter of the cable itself and the connectors. Here we discuss the cables from Figure 14–1 terminated with SMA connectors. The connector's outside diameter is 0.38 in. (9.7 mm). The outside diameter of the simplex cable is 0.189 in. (4.8 mm). The duplex cable has an oval cross section, 0.193 × 0.335 in. (4.9 × 8.5 mm). If the cable must be pulled through a conduit or duct, the minimum cross-sectional area required is 0.79 × 0.43 in. (20 mm × 11 mm) for duplex cable. For simplex cable, the minimum cross-sectional area is determined by the pulling grip used.

The primary consideration in selecting a route for fiber-optic cable through trays and ducts is to avoid potential cutting edges and sharp bends. Areas where particular caution must be taken are corners and exit slots in sides of trays (Figure 14–2).

If a fiber-optic cable is in the same tray or duct with very large, heavy electrical cables, care must be taken to avoid placing excessive crushing forces on the fiber-optic cable, particularly where the heavy cables cross over the fiber-optic cable (Figure 14–3). In general, cables in trays and ducts are not subjected to tensile forces; however, it must be kept in mind that in long vertical runs, the weight of the cable itself will create a tensile load of approximately 0.16 lb/ft (0.25 N/m) for simplex cable and 0.27 lb/ft (0.44 N/m) for duplex cable. This tensile load must be considered when determining the minimum bend radius at the top of the vertical run. Long vertical runs should be clamped at intermediate points (preferably every 1 to 2 m) to prevent excessive tensile loading on the cable. The absolute maximum distance between clamping points is 330 ft (100 m) for duplex cable and 690 ft (210 m) for simplex cable. Clamping forces should be no more than is necessary to prevent the possibility of slippage, and it is best determined experimentally since it is highly dependent on the type of clamping material used and the presence of surface contaminants, both on the clamp and on the jacket of the optical cable. Even so, the clamping force must not exceed 57 lb/in. (100 N/cm) and must be applied uniformly across the full width of the cable. The clamping force should be applied over as long a length of the cable as practical, and the clamping surfaces should be made of a soft material, such as rubber or plastic.

Tensile load during vertical installation is reduced by beginning at the top and running the cable down.

CONDUIT INSTALLATIONS

Fibers are pulled through conduits by a wire or synthetic rope attached to the cable. Any pulling forces must be applied to the cable members and not to the fiber. For cables without connectors, pull wire can be tied to Kevlar strength members, or a pulling grip can be taped to the cable jacket or sheath. More care, as discussed later in this chapter, must be exercised with cables with connectors.

SHARP BEND AT CORNER

FIGURE 14-2 Corners and exit slots (Courtesy of Canoga Data Systems)

FIGURE 14-3 Crossovers (Courtesy of Canoga Data Systems)

The first factor that must be considered in determining the suitability of a conduit for a fiber-optic cable is the clearance between the walls of the conduit and other cables that may be present. Sufficient clearance must be available to allow the fiber-optic cable to be pulled through without excessive friction or binding, since the maximum pulling force that can be used is 90 lb (400 N). Since minimum bend radius increases with increasing pulling force, bends in the conduit itself and any fittings through which the cable must be pulled should not require the cable to make a bend with a radius of less than 5.9 in. (150 mm). Fittings, in particular, should be checked carefully to insure they will not cause the cable to make sharp bends or be pressed against corners. If the conduit must make a 90° turn, a fitting, such as shown in Figure 14–4, must be used to allow the cable to be pulled in a straight line and to avoid sharp bends in the cable.

A pullbox is an access point in a conduit. Pullboxes should be used on straight runs at intervals of 250 to 300 ft to reduce the length of cable that must be pulled at any one time, thus reducing the pulling force. Also, pullboxes should be located in any area where the conduit makes several bends that total more than 180°. To guarantee that the cable will not be bent too tightly while pulling the slack into the pullbox, we must use a pullbox with an opening of a length equal to at least four times the minimum bend radius (4.75 in. or 120 mm). See Figure 14–5, which shows the shape of the cable as the last of the slack is pulled into the box. The tensile loading effect of vertical runs discussed in the section on tray and duct installations is also applicable to conduit installations. Since it is more difficult to properly clamp fiber-optic cables in a conduit than in a duct or tray, long vertical runs should be avoided, if possible. If a clamp is required, the best type is a fitting that grips the cable in a rubber ring. Also, the tensile load caused by the weight of the cable must be considered along with the pulling force to determine the maximum total tensile load being applied to the cable.

PULLING FIBER-OPTIC CABLES

Fiber-optic cables are pulled by using many of the same tools and techniques that are used in pulling wire cable. The departures from standard methods are due to the fact that the connectors are usually preinstalled on the cable, the smaller pulling forces that are allowed, and the minimum bend radius requirements.

The pull tape must be attached to the optical cable in such a way that the pulling forces are applied to the strength members of the cable (primarily the outer Kevlar layer), and the connectors are protected from damage. The recommended method of attaching a pulling tape to a simplex cable is the "Chinese finger trap" cable grip (Figure 14–6). The connector should be wrapped in a thin layer of foam rubber and inserted in a stiff plastic sleeve for protection. Since the smallest pulling grip available (Kellems 0333-02-044) is designed for 0.25-in.-(6.4-mm) diameter cable and the outside of the simplex cable is only 0.188 in.

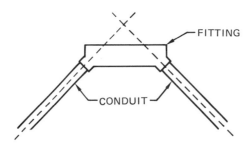

FIGURE 14-4   Turn fitting (Courtesy of Canoga Data Systems)

$R = 1.25$ in. (30 mm)

FIGURE 14-5 Pullbox dimensions (Courtesy of Canoga Data Systems)

FIGURE 14-6 Simplex pulling grip (Courtesy of Canoga Data Systems)

(4.8 mm), the cable grip should be stretched tightly and then wrapped tightly with electrical tape to provide a firm grip on the cable.

The duplex cable is supplied with Kevlar strength members extending beyond the outer jacket to provide a means of attaching the pulling tape (Figure 14-7). The Kevlar layer is epoxied to the outer jacket and inner layers to prevent inducing twisting forces while the cable is being pulled, since the Kevlar is wrapped around the inner jackets in a helical pattern. The free ends of the Kevlar fibers are inserted into a loop at the end of the pulling tape and then epoxied back to themselves. The connectors are protected by foam rubber and heat-shrink sleeving. The heat-shrink sleeving is clamped in front of the steel ring

FIGURE 14-7 Duplex pulling grip (Courtesy of Canoga Data Systems)

in the pulling tape to prevent pushing the connectors back toward the rest of the cable.

During pulling of the cable, pulling force should be constantly monitored by a mechanical gauge. If any increase in pulling force is noticed, pulling should immediately cease, and the cause of the increase should be determined. If the pulling wire is subject to friction, the tensile force on the pulling wire will be more than the force applied to the fiber-optic cable, resulting in false readings.

Pull tension can be monitored by a running line tensiometer or a dynamometer and pulley arrangement. If a power winch is used to assist the pulling, a power capstan with adjustable slip clutch is recommended. The clutch, set for the maximum loading, will disengage if the set load is reached. Relying on the experience of those pulling the cable is another alternative.

If necessary during pulling-in, the cable should be continuously lubricated. The lubricant can be poured or brushed on the cable as it enters the duct or conduit. Alternatively, a slit lubrication bag can be pulled ahead of the cable so that the lubricant spills out during the pull. Lubrication should be used only for difficult pulls.

At points such as pullboxes and manholes, where the cable enters the conduit at an angle, a pulley or wheel should be used to ensure that the cable does not scrape against the end of the conduit or make sharp bends. If the portion of the cable passing over the wheel is under tension, the wheel should be at least 12 in. (300 mm) in diameter. If the cable is not under tension, the wheel should be at least 4 in. (100 mm) in diameter.

As the cable emerges from intermediate-point pullboxes, it should be coiled in a figure-8 pattern with loops at least 1 ft in diameter for simplex and duplex cable and proportionally larger diameters for larger cables. The figure-8 prevents tangling or kinking of the cable. When all the cable is coiled and the next pull is to be started, the figure-8 coil can be turned over and the cable paid out from the top. This will eliminate twisting of the cable. The amount of cable that has to be pulled at a pullbox can be reduced by starting the pull at a pullbox as close as possible to the center of the run. Cable can then be pulled from one spool at one end of the run; then the remainder of the cable can be unspooled and coiled in a figure-8 pattern and pulled to the other end of the run.

Bends in the pull should be near the beginning of the pull. Pull forces are lower if the conduit bend is near the beginning. Bends tend to multiply, rather than add, tension. For example, if a cable goes in at 20 lb, it may come out of the bend at 30 lb. If it goes in at 200 lb, it may come out at 300 lb. In the first instance, the bend added 10 lb; in the second it added 100 lb.

SPLICE CLOSURES/ORGANIZERS

A *splice closure* is a standard piece of hardware in the telephone industry for protecting cable splices. You can see them along nearly every aerial telephone run, although they are also used in underground applications. Splices are protected mechanically and environmentally within the sealed closure. The body of the closure serves to join the outer sheaths of the two cables being joined.

FIGURE 14-8 Closure/organizer (Illustration courtesy of AMP Incorporated)

A standard universal telephone closure can also be used to house spliced fiber-optic cable. An *organizer panel* holds the splices. Most organizers contain provisions for securing cable strength members and for routing and securing fibers. Metal strength members may be grounded through the closure.

Organizers are specific to the splice they are designed to hold. In a typical application, the cable is removed to expose the fibers at the point it enters the closure. The length of fiber exposed is sufficient to loop it one or more times around the organizer. Such routing provides extra fiber for resplicing or rearrangement of splices. Figure 14-8 shows a closure housing fiber-optic organizers.

Closures can hold one or more organizers to accommodate 12 to 144 splices.

Closures have two main applications. The first is a straight splicing of two cables when an installation span is longer than can be accommodated by a single cable. The second is to switch between types of cables for various reasons. For example, a 48-fiber cable can be brought into one end of the closure. Three 12-fiber cables, all going to different locations, can be spliced to the first cable and brought out the other end of the closure.

DISTRIBUTION HARDWARE

Within a large fiber-optic system, signals must be routed to their final destination. Such routing involves the distribution signals and the fibers that carry them. The couplers described in Chapter 12 are one example. Here we describe hardware to allow transitions from one fiber type to another or distribution of fibers to different points.

In applications such as a building, a wiring center is used as a central distribution point. From this point, fibers can be routed to their destinations. A typical application example is an outdoor multifiber cable brought into the building to the distribution point. At the wiring center, each fiber in the outdoor cable is spliced to a simplex fiber that is routed to a different location, say a different floor.

FIGURE 14–9 Rack box/splicer organizer (Photos courtesy of ITT Electro-Optical Products)

A *rack box,* such as the example shown in Figure 14–9, containing an organizer similar to the organizer used in the splice closure, serves such distribution needs. The boxes mount in 19- and 23-in. equipment racks, which are standard sizes in the telephone and electronics industries. Besides the organizer, rack boxes contain provisions for securing fibers and strain-relieving cables as well as storage room for service loops.

PATCH PANELS

Patch panels provide a convenient way to rearrange fiber connections and circuits (Figure 14–10). A simple patch panel is a metal frame containing bushings in which fiber-optic connectors plug on either side. One side of the panel is usually "fixed," meaning that the fibers are not intended to be disconnected. On the other side of the panel, fibers can be connected and disconnected to arrange the circuits as required. Like rack boxes, patch panels are compatible with 19- and 23-in. equipment racks.

Patch panels are widely used in the telephone industry to connect circuits to transmission equipment. They can also be used in the wiring center of a building to rearrange connections. The splice organizer and the patch panel serve the similar function of distribution. The difference is that the organizer is intended for fixed, unchanged connections, whereas a patch panel is used for flexible, changeable connections.

FIGURE 14-10   Patch panel (Illustration courtesy of AMP Incorporated)

Some applications combine a splice organizer and patch panel. A large fiber is brought into the splice organizer and then spliced to other cables running directly into the "fixed" side of a patch panel. From the panel, cables can then be routed as necessary.

JUNCTION BOXES

The *junction boxes* described in this section can be thought of as being similar to electrical outlets. A home or office is wired for electrical power with cable running between walls or under floors. Electrical equipment plugs into this power at a junction box containing an electrical outlet.

Fiber-optic junction boxes serve a similar function except that they allow connection to cables carrying optical signals. In a building or office wired with fiber optics, the box serves as a transition point between the building cabling and the equipment. A short simplex or duplex cable, called a *jumper* or *drop cable*, runs from the junction box to the equipment being served.

Figure 14–11 shows a typical junction box, which mounts in a wall in the same manner as an electrical box. The cover plate contains two bushings for fiber-optic connectors. One side accepts the building wiring; the other side accepts the drop cable. When installed, the room side of the bushing faces down to prevent dust—a bane of any fiber-optic connection—from causing contamination when a drop cable is not connected.

FIGURE 14–11   Junction box (Illustration courtesy of AMP Incorporated)

FIGURE 14–12   Example of hardware applications

The distribution hardware clearly allows flexible, sophisticated application of fiber optics. It demonstrates that application of fiber-optic systems goes far beyond the simple point-to-point links emphasized in this book. The distribution hardware, pioneered originally for use in the telephone industry, has application outside of simple telephony. As we mentioned in Chapter 2, the information revolution has blurred the distinctions between simple telephone applications, telecommunications, and computers. Outside of telephone offices, the hardware finds use in wired buildings and offices, whether the application is telecommunications, computer networks, or local area networks. The couplers in Chapter 12 are used in LANs in an office environment, for example, whereas the distribution hardware discussed here might be used to connect several LANs within a building. Figure 14–12 shows a typical application showing the hardware discussed in the chapter.

SUMMARY

- The two most important concerns in installing a fiber-optic cable are minimum bend radius and maximum tensile load.
- Cables can be buried, strung aerially, or placed in trays and conduits.
- Closure/organizers are used to organize and protect spliced cable.
- Distribution boxes provide protection and organization for fibers and allow versatile and flexible distribution of fibers as application demands.
- Patch panels allow pluggable, rearrangeable interconnection of fibers.
- Junction boxes are similar to electrical outlets in that they allow equipment to be plugged into building wiring.

REVIEW QUESTIONS

1. What are the two most important factors to be considered when installing a cable?
2. Will a cable's minimum allowable bend radius be less during the installation or after?
3. Will a cable's maximum allowable tensile load be greater during the installation or after?
4. Name three additional influences against which a buried cable must be protected.
5. In an intrabuilding application, what distinguishes an ordinary duplex cable from a plenum duplex cable?
6. When pulling a cable through a conduit, to what part must the pulling forces be applied?
7. Why is a fiber-optic rack box either 19 in. or 23 in. wide?
8. What type of hardware allows connection between intrabuilding cable and drop cables from equipment?
9. What distinguishes a patch panel from a distribution box for splices?
10. If the total length of an application is twice the distance that can be achieved in a single pull, where is the best place to begin?

Chapter 15

Fiber-Optic Systems and Applications

This chapter looks at some typical applications using fiber optics. It is divided into three applications areas: computer, telecommunications, and networks. This division is for convenience, and there will be much overlap in actual applications. For example, a computer will be tied to a network, which in turn may be connected by a gateway to an outside telephone line. This whole system, in turn, could be in an industrial setting.

COMPUTERS

Two popular applications of fiber optics with computers are channel extenders and channel multiplexers.

Channel Extenders

Computers connect to other computers and to peripheral devices over a communications channel. Most personal computers and related equipment, for example, have an input/output channel variously called a serial port, com port, or RS-232C port. RS-232C is a specification recommending signal levels, circuit assignments, connector dimensions, and other matters to allow serial communication between electronic equipment. RS-232C serial communication is the most popular standard means of such communication.

Because the standard dates back to the early days of computer equipment, one of its largest drawbacks to many applications is relatively slow speed and limited transmission distances. At the highest data rate of 19.2 kbps, the recommended longest distance for transmission over copper wires is 15 m.

Figure 15–1 shows a fiber-optic RS-232C extender. The module contains a standard RS-232C connector at one end and fiber-optic transmitter and receiver ports at the other end. Internal to the module are the required electronic interface, transmitter, and receiver circuits. Transmitter output power is –27 dBm (2 μW) into a 50/125-μm fiber, and receiver sensitivity is –34 dBm (350 nW). Some RS-232C interfaces will supply enough power to operate the module, although

FIGURE 15-1   RS-232C transceiver (Courtesy of Canogo Data Systems)

the module has provisions for the required 12 Vdc from an external power supply.

The extender is transparent to the user. *Transparency* means that the user can connect a module to the equipment at each end without even realizing that fiber optics is involved. To the equipment at each end, the extender looks like a standard RS-232C channel. The difference is that transmission distances increase from 15 m to 1 km. Operating speeds are limited only by the speeds of the connected equipment, usually 0 to 19.2 kbps. Many extenders will operate faster, although the connected equipment may or may not. Similar extenders are available for most standard interfaces.

CHANNEL MULTIPLEXERS

Although an extender greatly increases transmission distances, it does not take advantage of the bandwidth of fiber optics. In some cases, the widely separated equipment will occur in clusters. A multiplexer allows several channels to be multiplexed over a fiber-optic cable. Full-duplex operation requires two fibers, one transmitting in each direction.

Figure 15-2 shows a 32-channel multiplexer, and Figure 15-3 shows a typical application. Again, the operation of the device is transparent, so no changes to the attached equipment are necessary. Each RS-232 channel can operate at 19.2 kbps.

Extended transmission distance is not the only reason for using fiber optics. Others are electrical isolation of equipment and elimination of ground loops.

FIGURE 15-2   RS-232C fiber-optic multiplexer (Courtesy of Canogo Data Systems)

FIGURE 15-3   Multiplexer application (Courtesy of Canoga Data Systems)

OPTICAL CHANNELS

The extenders and multiplexers use standard, existing interfaces and can be retrofitted on existing equipment. They are important in that they can be used with the existing base of equipment in use. Equipment is also available with optical interfaces built in. These interfaces are fiber optic from the start, not adaptations of electrical interfaces. The interfaces contain a transmitter/receiver pair for duplex operation. Of course, all interconnected equipment must have compatible optical interfaces. There are as yet no widely accepted standards for optical interfaces as there are for electrical interfaces. Each manufacturer, then, can design the interface differently, so we often cannot connect equipment from one manufacturer to equipment from another manufacturer. Although standard electrical interfaces such as RS-232C can be specified for most electronic equipment, it cannot be done for optical interfaces. Even so, such standards will evolve as fiber optics becomes more widely used in equipment.

TELECOMMUNICATIONS

The telephone industry is the heaviest user of fiber optics because of its high bandwidth, low losses, and electrical immunity. The two most commonly used fibers are 50/125 multimode and single mode.

The earliest uses of fiber optics were trunk lines connecting central offices and long-distance toll centers. The central office contains the switching equipment providing telephone service for a given geographical area. The central office is designated by the first three numbers of a seven-digit telephone number. Toll centers contain switching equipment for distributing calls from one metropolitan center to another.

Fiber optics is also finding application in the subscriber loop, which is the circuit connecting a central office to subscriber telephones. From the central office, feeder lines connect to a smaller electronic system called the *serving area interface*. From this interface go the lines that connect directly into the telephones in homes. At present, fiber optics is used on the feeder lines between the central office and the serving area interface. The connection from the interface to subscribers is still by copper wire, although this situation should change rapidly as the uses of fiber optics continue to expand. Figure 15–4 shows a simplified diagram of a telephone network.

FIGURE 15–4   Telephone system (Courtesy of Corning Glass Works)

90-MBPS FIBER-OPTIC SYSTEM EXAMPLE

Figure 15–5 is a photograph of the FTS-3C fiber-optic transmission system offered by ITT Telecom. It is a 90-Mbps system that is capable of transmitting and receiving 1344 channels on a pair of optical fibers. The different subsystems of the FTS-3C plug into the shelf (or frame) of the unit. The shelf mounts in a 19-in. relay rack, which is a standard rack in the telephone industry.

Figure 15–6 shows a simple block diagram of a typical FTS-3C application. Twenty-eight DS1 signals are fed to each of two M13 multiplexers, for a total of 56 DS1 signals into two multiplexers. A DS1 signal is a basic telephone signaling rate providing 24 voice channels at a data rate of 1.544 Mbps, which is the basic digital data rate. Most faster, higher-capacity systems are simply multiples of the DS1 signal. The 45-Mbps telephone system uses 28 DS1 signals, and the 90-Mbps system uses 56.

Each multiplexer combines (multiplexes) the 28 individual DS1 signals into a single DS3 signal. A DS3 signal is another standard data rate, equal to 44.736 Mbps or 672 voice channels. The first fiber-optic telephone system, in 1977, operated at this rate.

The CPP unit divides the DS3 signal from the multiplexer into two identical DS3 signals: One is used for transmission; the other, used for a protection system, is available for monitoring, testing, or other purposes.

Each DS3 signal is then fed to the FTS-3C fiber-optic system. The FTS-3C combines these into a single 90-Mbps data stream and converts the data to NRZ

FIGURE 15–5 FTS-3C 90-Mbps fiber-optic transmission system (Photos courtesy of ITT Electro-Optical Products)

FIGURE 15-6   Application of FTS-3C (Photos courtesy of ITT Electro-Optical Products)

format for transmission. At the same time, it adds additional parity and framing bits, so the actual optical signal is 100.8 Mbps. This adding of additional bits to the signals is called *stuffing.*

The optical transmitter takes this 100.8 Mbps data stream and converts it into an optical signal and couples it into a fiber for transmission. At the other end, another FTS-3C system receives the signal. The receiver system converts the optical signal back into a 100.8 Mbps electrical signal with recovered clock. The signal is then divided into two separate 50.4-Mbps data channels. The stuffed parity and framing bits are then removed, reducing the data rate of each channel from 50.4 to 44.736 Mbps. The data streams are then reconverted from an NRZ format to the B3ZS format used by the telephone industry. Each M13 multiplexer then receives one of the DS3 signals and demultiplexes it back into 28 DS1 channels.

FIGURE 15-7   45-Mbps transmitter unit (Photos courtesy of ITT Electro-Optical Products)

The transmitter unit consists of a laser source and the laser drive circuitry. The circuitry includes a pin photodiode that monitors part of the laser's output. The pin diode produces a current proportional to the optical output power. This current is used to control the bias current to the laser to maintain constant output levels under varying temperature conditions and laser aging. The circuit also contains an alarm that signals that an aging laser has about 1000 hr of life left.

The transmitter is available for either 850-nm short-wavelength or 1300-nm long-wavelength operation. The laser launches about 1 mW (0 dBm) of optical power into a 50/125 fiber. Figure 15–7 shows the details of a typical transmitter unit, in this case a 45-Mbps unit.

The receiver uses an APD to detect the optical power and convert it to an electrical current. It is available with a dynamic range of either –8 to –46 dBm or –12 to –38 dBm.

Figure 15–8 shows the flow of signals through the FTS-3C system.

LOCAL AREA NETWORKS

The discussion of couplers in Chapter 12 introduced LANs, which are multi-terminal communication networks in a limited geographical area. It illustrated three popular topologies for LANs: bus, ring, and star.

Besides topology, LANs are distinguished by

- Signaling technique
- Transmission medium
- Access method

Signaling Technique

The signaling technique used in a LAN can be either baseband or broadband. In baseband communication, the channel is occupied by a single signal, usually without a carrier. On/off pulses of digital communication is an example. In broadband communication, several signals occupy the channel. The signals are usually placed on a carrier. For example, cable television uses broadband methods and frequency-division multiplexing. Each separate television channel is placed on a separate carrier frequency.

Fiber-optic communication is somewhat confusing insofar as baseband and broadband communication, several signals occupy the channel. The signals are usually placed on a carrier. For example, cable television uses broadband meth-optics is better considered baseband, since it more closely resembles baseband communication. On/off pulses of voltage sent on a wire is baseband communication, and turning a source on and off to produce optical pulses is essentially the same. Sending several optical channels through WDM is a broadband technique.

Transmission Medium

The three most widely used transmission media are twisted-pair wires, coaxial cable, and fiber-optic cable.

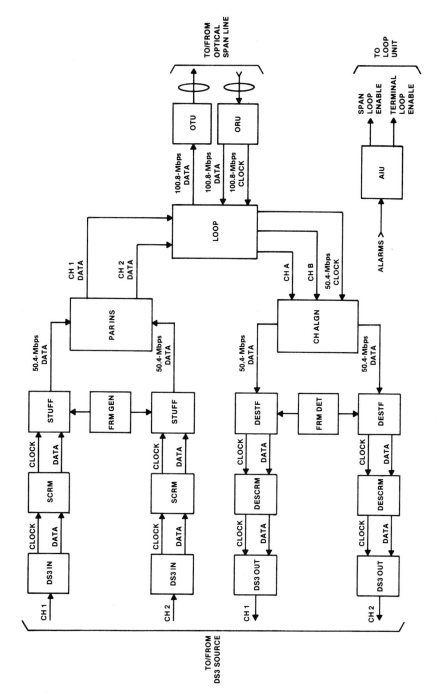

FIGURE 15–8   Signal flow in FTS-3C (Photo courtesy of ITT Electro-Optical Products)

Access Method

Access refers to the method by which a terminal gains control of the network to send messages. Two popular methods are carrier sense multiple access with collision detection (CSMA/CD) and token passing.

In CSMA/CD, every terminal has equal access to the network at any time (multiple access). Before seizing access, a station first listens to the network to sense if another station is already transmitting (carrier sense). If the station senses another message on the network, it does not gain access. It waits awhile and listens again for an empty network.

The possibility still exists that two terminals will listen and sense an empty network. Each will place its message on the network, where the messages will collide and become garbled. Therefore, collision detection is also necessary. Once a collision is detected, the garbled messages will be ignored. The terminals will begin again and listen for an empty network.

Bus networks use CSMA/CD access methods.

In a token passing network, a token is passed from terminal to terminal around the network. Only when it has the token is a terminal allowed access to a network. Ring networks often use token passing.

ETHERNET SYSTEM

The Ethernet LAN is a bus network using CSMA/CD. Originally created by Xerox Corporation, Digital Equipment Corporation, and Intel Corporation, Ethernet was designed to use coaxial cable at data rates up to 10 Mbps. Beyond topology, access method, signaling technique, and media, the Ethernet specification also details the format for messages so that each station can originate and receive messages that will be understood by all other stations on the network.

Fiber optics has also been used to create an Ethernet-compatible system. Figure 15–9 shows an example of such a network.

In Chapter 12, we saw that a bus network uses tee couplers. The drawback of a tee coupler is that it is usable with only a limited number of terminals. A star coupler, on the other hand, supports a greater number of terminals. The network in Figure 15–9 uses a transmissive star coupler for this reason. Thus, the physical topology of the network is a star, but the logical topology of the network is a bus.

The star wiring center (Figure 15–10) contains the star coupler and the collision-detection electronics. To detect collisions, each incoming fiber to the star first connects to an optical tap, a tee coupler in which only a small percentage of the light is tapped. Each tap connects to a receiver whose sole function is to detect optical power. When more than one of the receivers detects power, a collision has occurred. The collision-detection electronics sends a jam signal to all transceivers. One of the input ports to the star is used for the jam signal.

Each terminal connects to a fiber-optic transceiver through a *network access unit* (NAU). The NAU is an electronic circuit for connecting a terminal to an

FIGURE 15-9   Fiber-optic Ethernet network (Courtesy of Siecor Corporation)

FIGURE 15-10   Star wiring center (Courtesy of Siecor Corporation)

Ethernet transceiver, be it electronic or optical. Any electronic equipment containing an Ethernet NAU can attach to either a coaxial or an optical network.

Each transceiver has a transmitter power of –9 dBm and a receiver sensitivity of –36 dB, yielding a system power budget of 27 dB. The transceiver power levels take into account loss from the SMA connectors used to connect the cable to the transmitter and receiver ports. The coupler uses fiber pigtails, and the recommended method of connecting network fibers to the star is by an elastomeric splice. Since the maximum loss of a splice at each of the required two splices (one input, one output) is 0.5 dB, the loss for connecting each transceiver/terminal to the coupler is 1 dB. The cable used is a 100/140-μm fiber with a loss of 6 dB/km.

The number of ports in the star coupler determines the number of terminals in the network. The coupler also determines the maximum cable length of each leg of the network. System loss is divided between the fiber-optic cable and the coupler. Since coupler insertion loss increases with the number of ports, the maximum cable length decreases as the size of the network increases. Table 15–1 summarizes the number of terminals, star coupler requirements, and maximum cable lengths. Remember that the collision-detection electronics uses one or two of the input ports of the star. The coupler loss includes the loss optical tap to the collision-detection electronics. Unused ports do not affect network operation. We may, for example, construct a 10-terminal network with a 32 × 32 coupler so that the network can be expanded in the future without changing the star wiring center.

The maximum cable length is the distance separating the two most distant transceivers in the network, not the distance from a transceiver to the star

STATIONS	COUPLER	COUPLER LOSS (dB)	CABLE LENGTH (km)
3	4 × 4	8	2.5
7	8 × 8	12	1.8
14	16 × 16	15	1.3
30	32 × 32	18.5	0.75
62	64 × 64	22	0.33

TABLE 15–1   Summary of coupler loss and cable length of different sized networks

Transmitter power	–9 dBm (125 μW)	
Receiver sensitivity	–36 dBm (250 μW)	
Power budget		27 dB
Two splices	1 dB	
1 km of cable/leg	6 dB	
8 × 8 star coupler	12 dB	
Total loss		19 dB
Power margin		8 dB

FIGURE 15–11   Typical network power budget

coupler. For the 2.5-km cable length, for example, one terminal might be 1 km from the star coupler, and the second terminal 1.5 km away. In a copper-based Ethernet system, the maximum bus length is 500.

Figure 15–11 shows the typical power budget.

The 8-dB margin applies to each leg of the network. The network can be extended to allow longer leg lengths to each terminal or more terminals beyond 62 by using an optical repeater.

SUMMARY

- Two applications of fiber optics in computers are channel extenders and channel multiplexers.
- A one-channel extender extends the transmission distance over that of a standard copper-based communication channel such as RS-232C.
- A channel multiplexer combines several communication channels over a single optical fiber.
- An FTS-3C fiber-optic system transmits at 90 Mbps.
- A basic digital telephone data rate is DS1, which is equal to 1.544 Mbps or 24 voice channels. Higher digital data rates are usually multiples of the basic DS1 rate.
- LANs use either baseband or broadband signaling techniques. Fiber optics is best considered baseband.
- Access method in a LAN is the method by which a terminal gains control of the network to send messages.
- Two popular access methods are token passing and CSMA/CD.
- The Ethernet LAN is a bus network using CSMA/CD.
- The star wiring center of a fiber-optic Ethernet contains a star coupler and collision detection electronics.

REVIEW QUESTIONS

1. Define a channel extender and give an example of its use.
2. Does an RS-232C port use serial or parallel communication?
3. What is meant by user transparency?
4. What are the two most commonly used fiber sizes in telecommunications?
5. How many DS1 signals does a FTS-3C telephone system use? How many voice channels is that?
6. Define bit stuffing.
7. What is the importance to a LAN of carrier sense and collision detection?
8. How does an Ethernet LAN detect collisions?
9. What is the cable size most commonly used in a fiber-optic LAN? Why is this size preferred over the sizes used in telecommunication?
10. What determines the maximum cable of each leg in a network?

Chapter 16

Introduction to Test and Other Equipment

This chapter provides a brief look at some of the equipment commonly used in the field to install, inspect, and maintain fiber-optic systems. It also discusses methods to measure fiber attenuation and connector insertion loss. The equipment discussed is the optical power meter, optical time-domain reflectometer, fusion splicer, inspection microscope, polishing machine, and a few hand tools. Our discussion will be simplified and clarified by looking at specific examples of each.

FIBER-OPTIC TESTING

A great many tests must be performed on optical fibers. A fiber manufacturer must test a fiber to determine the characteristics by which the fiber will be specified. As a quality control measure during manufacture of fibers, the manufacturer must constantly test the fibers to ensure that they meet the specifications. Among such tests are the following:

- Core diameter
- Cladding diameter
- NA
- Attenuation
- Refractive index profile
- Tensile strength

Other tests performed on fibers or on fiber-optic cables concern their mechanical and environment characteristics. Mechanical tests such as impact resistance, tensile loading, and crush resistance test the cable's ability to withstand physical and mechanical stresses. Environmental tests evaluate the changes in attenuation under extremes of temperature, repeated changes (cycling) of temperature, and humidity.

Fiber and cables are specified as the result of such tests. Engineers pick cables suited to their applications on the basis of specifications. Manufacturers of other fiber-optic components—sources, detectors, connectors, and couplers—also test to characterize and specify their offerings.

Other tests are performed during and after installation of a fiber-optic link. These tests ensure that the installed system will meet performance requirements. Splices, for example, must not exceed certain loss values. During installation, each splice must be tested. If the loss is unacceptably high, the fibers must be respliced. This chapter describes how such a test is done with optical time reflectometry.

OPTICAL POWER METER

The *optical power meter* is analogous to the volt-ohm-milliammeter (VOM) used in electronics. Figure 16-1 shows a power meter. The meter reads optical power levels, either in absolute levels of dBm and dBμ or in relative units of dB.

The meter itself is completely electronic. Sensors that plug into the unit contain the detector and perform the optical-to-electrical conversion. Different sensors are available for use at different power levels and operating wavelengths from 400 to 1800 nm. Adapters permit bare fibers or a variety of popular connectors to be connected to a sensor.

The range of the meter, with different sensor heads, is –80 dBm (10 pW) to +33 dBm (2 W). Resolution is selectable at 0.1 or 0.01 dB. Resolution determines whether the reading will be one or two decimal places. An output jack, providing an adjustable, regulated current, allows plug-in LEDs to provide the light source for measurements.

The meter can be used for a number of measurements. We will describe its use for fiber attenuation measurements and connector insertion loss measure-

FIGURE 16-1   Optical power meter (Courtesy of Photodyne)

ments. But first, we must discuss mode control, an important factor to test results and repeatable tests.

MODE CONTROL

Chapter 11 discussed the influence of launch and receive conditions on connector insertion loss. Such launch and receive conditions result from the mode distribution in fibers. Long launch or receive conditions mean that the equilibrium mode distribution exists in the fiber. For repeatable measurements, launch conditions especially must be controlled. Most often, long launch, or EMD, conditions are preferred. The Electronics Industry Association (EIA) recommends a 70/70 launch: 70% of the fiber core diameter and 70% of the fiber NA should be filled. This recommendation corresponds to EMD in a graded-index fiber. EMD can be reached by three approaches: optical, filtering, and long fiber length.

The optical approach uses collimating and focusing lenses and optical apertures to direct the light into the center of the fiber core. Although the result is precise, this approach is only practical in a laboratory equipped with an optics bench.

The long-fiber approach uses a fiber long enough to reach EMD regardless of the light injected into it. Even though the source overfills or underfills the fiber, EMD is reached over distance. The disadvantage is that the required fiber length may be several kilometers.

Filtering presents the most practical approach. A mode filter causes mode mixing to simulate EMD in a short length of fiber. A $\frac{1}{2}$ in. diameter mandrel around which a fiber is wrapped five or six times is a standard approach. The five wraps cause mode mixing so that high-order modes in an overfilled fiber are lost.

The mandrel wrap, however, does not remove cladding modes. Cladding modes will propagate a short distance in multimode fiber, and it may be necessary to remove them if the fiber being used is short. A cladding mode stripper is a material with a refractive index greater than the cladding. A bath of glycerin, oil, or other suitable liquid will serve. The buffer coating must be removed to expose the cladding, which is immersed in the mode stripper. Because of the greater refractive index of the stripper, light is no longer bound in the cladding. It passes into the stripper.

FIBER LOSS MEASUREMENTS

Losses in a fiber-optic cable can be measured in two ways. The cutback method, shown in Figure 16–2, uses a single fiber. First, power through the fiber is measured. A section of fiber is then cut off, and the power through the remaining fiber is measured. The fiber loss in decibels per kilometer can then be determined by

$$\text{loss} = \frac{P_2 - P_1}{L}$$

where P_1 is the dBm reading from the first measurement, P_2 is the dBm reading from the second measurement, and L is the difference, in kilometers, in the two cable lengths. Notice that subtracting one dBm unit from another yields dB. Changing the units of the difference in fiber lengths allows loss per unit length to be measured in other units.

A second method of measuring attenuation in an optical fiber is to compare the power through the cable under test to the power of a known reference cable. The power through the cable under test is measured in absolute units on the meter. Next, power through the reference cable is measured.

A power meter can also be used to check the output power from a transmitter during quality control or troubleshooting.

INSERTION LOSS TESTS

Chapter 11 discussed insertion loss testing for connectors and splices briefly. In essence, the test measures the power through a length of fiber. The fiber is then cut near the center, and the connector or splice is applied to the exposed ends. Power through the cable is again measured. The difference between the two readings is the insertion loss contributed by the connector or splice. Here we wish to emphasize the importance of mode control in the test. Since losses will vary, depending on the launch and receive conditions, controlling these conditions with mode filters and cladding mode strippers is essential to understandable, meaningful, and repeatable results.

A main drawback to the use of an optical power meter in many applications is that both ends of the fiber must be available. An optical time-domain reflectometer allows testing when only one end of the fiber is conveniently available.

FIGURE 16–2 Attenuation measurement by cutback technique (Courtesy of Photodyne)

TIME AND FREQUENCY DOMAINS

A signal can be described in terms of the time domain or the frequency domain. This book has used both domains without bothering to distinguish between them. We described the limitations on a fiber's information-carrying capacity in terms of both bandwidth and dispersion. Bandwidth is in the frequency domain; rise-time dispersion is in the time domain. Analog engineers, dealing with analog signals in the frequency domain, talk in terms of frequencies. Digital engineers, dealing with pulses, use the time domain and talk in terms of rise times and pulse widths. We have given a simple method of relating the frequency and time domain in the equations relating bandwidth and rise time.

$$t_r = \frac{0.35}{BW}$$

$$BW = \frac{0.35}{t_r}$$

OPTICAL TIME-DOMAIN REFLECTOMETRY

As its name implies, the *optical time-domain reflectometer* allows evaluation of an optical fiber in the time domain. Figure 16–3 shows an optical time-domain reflectometer. The following paragraphs describe the principles of optical time-domain reflectometry (OTDR).

OTDR relies on the backscattering of light that occurs in an optical fiber. Some of the light entering the fiber will be reflected back to the light. *Backscattered* light is that light that reaches the input end of a fiber. Backscattering results from Rayleigh scattering and Fresnel reflections. Rayleigh scattering, remember, is scattering caused by the refractive displacement due to density and compositional variations in the fiber. In a quality fiber, the scattered light can be assumed to be evenly distributed with length. Fresnel reflections occur because of changes in refractive index at connections, splices, and fiber ends. A portion of the Rayleigh scattered light and Fresnel reflected light reaches the input end as backscattered light.

Figure 16–4 shows a simple block diagram of an OTDR unit. Its main parts are a light source, a beamsplitter, a photodetector, and an oscilloscope. A short, high-powered pulse is injected through the beamsplitter into the fiber. This light is then backscattered as it travels through the fiber. The beamsplitter directs the backscattered light to the photodetector. The amplified output of the detector serves as the vertical input to the oscilloscope. Because the power to the detector is extremely small, repeated measurements are made by the electronics of the ODTR; the SNR is improved by averaging the readings, after which the results are displayed.

The OTDR screen displays time horizontally and power vertically. Fiber attenuation appears as a line decreasing from left (the input end of the fiber) to

(A)

(B)

FIGURE 16–3 Optical time-domain reflectometer (A) and close-up of screen (B) (Courtesy of Anritsu America, Inc.)

FIGURE 16–4 OTDR block diagram (Courtesy of Photodyne)

right (the output end). Both the input and the backscattered light attenuate over distance, so the detected signal becomes smaller over time. A connector, splice, fiber end, or abnormality in the fiber appears as an increase in power on the screen, since backscattering from Fresnel reflections will be greater than backscattering from Rayleigh scattering. The quality of a splice can be evaluated by the amount of backscattering: Greater backscattering means a higher-loss splice. A connector shows both a power increase from reflection and a power drop from loss. The degree of loss indicates the quality of the connection. Figure 16–5 shows a typical display. Notice that it is similar to the power budget diagram in Chapter 13.

Light travels through a fiber at a speed of about 5 ns/m, depending on the refractive index of the core. Time can be correlated to distance by

$$D = \frac{ct}{2n}$$

where D is distance along the fiber, c is the speed of light, t is the round-trip travel time of the input pulse, and n is the core's average refractive index. Most OTDRs use a cursor to mark a horizontal point on the trace and display the distance in terms of both time and physical distance. One can, for example, determine the distance to a splice with a great degree of accuracy, typically to within a foot. The OTDR measures the distance along the fiber, not necessarily the length of the cable, however. If the fiber is stranded around a central core, the actual fiber length will be somewhat longer than the cable length.

The lengths at which the OTDR can be used depend on two things. First is the dynamic range of the units, which sets the minimum and maximum power to the detector. Beyond that, the length is determined by fiber attenuation and losses at splices and connectors. OTDR dynamic range and power losses within the fiber system set the length possible before the backscattered light becomes too weak to be detected. With typical low-loss fibers used in long-distance telecommunications, the OTDR can be used on lengths of 20 to 40 km.

USES OF OTDR

OTDR has many uses in research and development and in manufacturing. Here we will discuss three field uses: loss per unit length, splice and connector evaluation, and fault location. These three uses are very important during installation and for system maintenance. Loss per unit length and evaluation of connectors and splices can also be done with an optical power meter. The advantage of OTDR in field applications involving any appreciable length is that only one end of the fiber must be available. The OTDR unit can be placed in a sheltered environment and kept out of manholes. In applications involving short lengths of fibers, such as offices, the power meter might be more convenient.

Loss Per Unit Length

The loss budget of a fiber-optic installation assumes certain fiber attenuation, which is loss per unit length. OTDR can measure attenuation before and after installation. Measurements before installation ensure that all fibers meet specified limits. Measurements after installation check for any increases in attenuation that could result from such things as bends or unexpected loads.

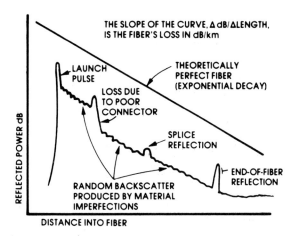

FIGURE 16-5 OTDR display (Courtesy of Photodyne)

Splice and Connector Evaluation

Here OTDR can be used during the installation processes to ensure that all splices and connector losses are within acceptable limits. After the splice or connector is applied, an OTDR reading checks the loss. If it is unacceptable, another connector or splice is applied.

Some splices are tunable. By rotating one half of the splice in relation to the other, losses from lack of symmetry in the fiber or splice can be minimized. As the splice is rotated, the OTDR display is monitored to determine the point of lowest loss and maximum transmission.

The OTDR unit and operator are usually in a location such as a central telephone office. Communication with the splicer is by radio or telephone.

Fault Location

Faults, such as broken fibers or splices, may occur during or after installation. A telephone line will be broken two or three times during its 30-year lifetime. The location of the fault may not be apparent if the cable is buried or in a conduit or if only the fiber and not the cable is broken. OTDR provides a useful method of locating the fault accurately and thereby saving great time and expense. It certainly beats digging up several kilometers of cable unnecessarily.

FUSION SPLICER

Chapter 11 described a fusion splice. Here we expand by looking at a splicer for multimode fibers. The fusion splicer shown in Figure 16-6 is held in a rugged aluminum case, so it is easily carried down manholes or up telephone poles. It is powered by a lead-acid battery, which permits about 150 splices to be made on a single charge.

FIGURE 16-6 Fusion splicer (Courtesy of Siecor Corporation)

A 20-kHz power supply produces an adjustable 0-to-20-mA current for the electric arc. Fusion temperatures reach about 2000°C. A prefusion arc cleans the fibers and rounds their end, and the fusion arc joins them. The arc current is adjustable from 0 to 20 mA, and the timing of each arc can be set independently from 0 to 0.09 s in 0.1-s steps. The arc produces a high surface tension on the molten glass that results in self-alignment of the fibers. Misalignments of up to 10 μm are automatically compensated for without significantly adding to attenuation.

To produce good, highly perpendicular cleaves, the unit has a cleaving tool that uses a stress bend and carbide blade. After cleaving, the fibers are placed in guides that automatically align them. The high precision of the guides and the self-aligning properties of the electric arc make axial adjustment of the fiber the only adjustment necessary. Micropositioners allow three-dimensional alignment of each fiber along x, y, and z axes. Fibers can also be rotated to increase the parallelism between nonperpendicular end faces.

A 50X microscope allows the alignment to be observed. Alignment marks in the viewing area of the microscope aid positioning. An optional automatic feeder automates the fusion process. After the fibers have been aligned with the microscope marks, pressing a button automatically performs prefusion, feeding of the fibers closer together, and fusion.

The splicer can also hold a splice organizer such as used to hold splices in closures and rack boxes. The ability to do so greatly simplifies the splicing procedure, since fibers can be routed as they are spliced. Finished splices can be checked by OTDR for losses.

POLISHING MACHINE

The polishing machine shown in Figure 16-7 partially automates the polishing process to produce consistent end finishes. The machine contains two polish-

FIGURE 16-7 Fiber-optic polishing machine (Courtesy of Buehler Ltd., 41 Waukegan Road, Lake Bluff, IL 60044)

ing platens: one for rough polishing and one for final, fine polishing. Replaceable circular polishing disks adhere to the platens.

Connectors held in precision adapters attach to the sweep arm. The arm moves up and down at 12 sweeps per minute while the platens spin. The adapters contain jeweled stops that precisely control the polishing depth.

Because connectors are polished, the fibers are not often carefully cleaved flush against the connector face. As a result, the fiber often protrudes from the connector. The drive mechanism uses a decreasing pneumatic pressure to make soft starts during polishes. The soft start prevents damage to the polishing disk and, more importantly, fractures to the fiber. Polishing pressure is then increased. Approximate polishing times are 30 s for rough and 45 s for fine. After polishing, the finish should be inspected under a microscope or magnifier.

INSPECTION MICROSCOPE

Portable inspection microscopes (Figure 16–8) allow closeup inspection of cleaved fiber ends and polishing finishes in the field. Closeup inspection is an important quality control procedure for evaluating splices and connectors. It is much simpler than OTDR, although it can be used in conjunction with OTDR.

The microscope allows both customary through-lens illumination and oblique illumination that reveals pits, scratches, and polishing patterns. Connectors or fibers can be rotated to allow finishes to be viewed across the end, straight on, or at angles in between. Different adapters are required for different connectors and bare fibers. Interchangeable eyepiece lenses and objective lenses allow a magnification range from 40X to 800X.

FIGURE 16-8   Inspection microscope (Illustration courtesy of AMP Incorporated)

INSTALLATION KITS

Installation of fiber-optic cables requires many tools, some of which are common and some of which are special to fiber optics and the products being used. An installation kit, such as shown in Figure 16-9, contains all the tools commonly needed to split open multifiber cables, prepare fibers, apply connectors, polish connectors, and inspect the finish.

A partial list of the kit contents includes

- Screwdriver and nutdriver for preparing organizers
- Side-cutting pliers and utility knife for removing cable sheaths
- Tape measure for measuring cable and fiber stripping lengths
- Cable stripper for removing jacketing

FIGURE 16-9   Installation kit (Illustration courtesy of AMP Incorporated)

- Fiber strippers for removing buffer coatings
- Scissors for trimming Kevlar strength members
- Roller for mixing epoxy
- Scribe tool for trimming fibers
- Crimp tool for crimping connectors to fibers
- Polishing materials
- Duster (a canned air for cleaning fibers)
- Heat gun for applying heat-shrink tubing
- Index-matching fluid for lowering or eliminating loss from Fresnel reflections in a fiber-to-fiber interconnection
- Inspection microscope for inspecting fiber end finishes

Such kits provide the proper tools for the job in one convenient location.

SUMMARY

- An optical power meter allows optical power levels to be measured.
- Mode control is often important to repeatable test results.
- An optical time reflectometer is used to evaluate connectors and splices, locate faults, and measure loss per unit length in a fiber.
- A fusion splicer fuses fiber in fiber-to-fiber interconnections.
- A microscope allows end finishes to be inspected.
- A polishing machine produces fine, consistent end finishes.

REVIEW QUESTIONS

1. Name the electrical test equipment to which an optical power meter is most easily compared.
2. Describe a method of measuring fiber loss with a power meter.
3. List three methods of obtaining EMD in a fiber under test.
4. For each of the following, say whether the time domain or the frequency domain applies:
   A. Rise time
   B. Bandwidth
   C. Pulse width
   D. 500 MHz
   E. 20 ns
   F. Dispersion
   G. Source spectral width
   H. ODTR
5. Name two factors that limit the distance at which an OTDR unit can be used.
6. List three uses of OTDR.
7. Sketch an OTDR display for a 20-km link with a splice at 5 km and connectors at 10 and 15 km.

8. Beside performing the splice itself, name two other essential activities required of a good splice that the fusion splicer described in this chapter allows to be performed.
9. How many polishing stages are there in polishing a typical connector? What is the function of each stage?

Glossary

Absorption   Loss of power in an optical fiber, resulting from conversion of optical power into heat and caused principally by impurities, such as transition metals and hydroxyl ions, and also by exposure to nuclear radiation.

Acceptance angle   The half-angle of the cone within which incident light is totally internally reflected by the fiber core. It is equal to arcsin (NA).

AM   Amplitude modulation.

Amplitude modulation   A transmission technique in which the amplitude of the carrier is varied in accordance with the signal.

Angular misalignment   The loss of optical power caused by deviation from optimum alignment of fiber to fiber or fiber to waveguide.

APD   Avalanche photodiode.

Attenuation   A general term indicating a decrease in power from one point to another. In optical fibers, it is measured in decibels per kilometer at a specified wavelength.

Attenuation-limited operation   The condition in a fiber-optic link when operation is limited by the power of the received signal (rather than by bandwidth or by distortion).

Avalanche photodiode   A photodiode that exhibits internal amplification of photocurrent through avalanche multiplication of carriers in the junction region.

Backscattering   The return of a portion of scattered light to the input end of a fiber; the scattering of light in the direction opposite to its original propagation.

Bandwidth   A range of frequencies.

Bandwidth-limited operation   The condition in a fiber-optic link when bandwidth, rather than received optical power, limits performance. This condition is reached when the signal becomes distorted, principally by dispersion, beyond specified limits.

Baseband   A method of communication in which a signal is transmitted at its original frequency without being impressed on a carrier.

Baud   A unit of signaling speed equal to the number of signal symbols per second, which may or may not be equal to the data rate in bits per second.

Beamsplitter   An optical device, such as a partially reflecting mirror, that splits a beam of light into two or more beams and that can be used in fiber optics for directional couplers.

Bend loss   A form of increased attenuation in a fiber that results from bending a fiber around a restrictive curvature (a macrobend) or from minute distortions in the fiber (microbends).

BER   Bit-error rate.

Bit   A binary digit, the smallest element of information in binary system. A 1 or 0 of binary data.

Bit-error rate   The ratio of incorrectly transmitted bits to correctly transmitted bits.

Broadband   A method of communication in which the signal is transmitted by being impressed on a higher-frequency carrier.

Buffer   A protective layer over the fiber, such as a coating, an inner jacket, or a hard tube.

Buffer coating   A protective layer, such as an acrylic polymer, applied over the fiber cladding for protective purposes.

Buffer tube   A hard plastic tube, having an inside diameter several times that of a fiber, that holds one or more fibers.

Bus network   A network topology in which all terminals are attached to a transmission medium serving as a bus.

Byte   A unit of 8 bits.

Carrier sense multiple access with collision detection   A technique used to control the transmission channel of a local area network to ensure that there is no conflict between terminals that wish to transmit.

Channel   A communications path or the signal sent over that channel. Through multiplexing several channels, voice channels can be transmitted over an optical channel.

Cladding   The outer concentric layer that surrounds the fiber core and has a lower index of refraction.

Cladding mode   A mode confined to the cladding; a light ray that propagates in the cladding.

Connector   A device for making connectable/disconnectable connections of a fiber to another fiber, source, detector, or other devices.

Core   The central, light-carrying part of an optical fiber; it has an index of refraction higher than that of the surrounding cladding.

Coupler   A multiport device used to distribute optical power.

CSMA/CD   Carrier sense multiple access with collision detection.

Cutoff wavelength   For a single-mode fiber, the wavelength above which the fiber exhibits single-mode operation.

Dark current   The thermally induced current that exists in a photodiode in the absence of incident optical power.

Data rate   The number of bits of information in a transmission system, expressed in bits per second (bps), and which may or may not be equal to the signal or baud rate.

dB   Decibel.

dBm   Decibel referenced to a milliwatt.

dBμ Decibel referenced to a microwatt.

Decibel   A standard logarithmic unit for the ratio of two powers, voltages, or currents. In fiber optics, the ratio is power.

$$dB = 10 \log_{10} \left(\frac{P_1}{P_2} \right)$$

Detector   An optoelectronic transducer used in fiber optics for converting optical power to electric current. In fiber optics, usually a photodiode.

Diameter-mismatch loss   The loss of power at a joint that occurs when the transmitting half has a diameter greater than the diameter of the receiving half. The loss occurs when coupling light from a source to fiber, from fiber to fiber, or from fiber to detector.

Dichroic filter   An optical filter that transmits light selectively according to wavelength.

Diffraction grating   An array of fine, parallel, equally spaced reflecting or transmitting lines that mutually enhance the effects of diffraction to concentrate the diffracted light in a few directions determined by the spacing of the lines and by the wavelength of the light.

Dispersion   A general term for those phenomena that cause a broadening or spreading of light as it propagates through an optical fiber. The three types are modal, material, and waveguide.

Distortion-limited operation   Generally synonymous with bandwidth-limited operation.

Duplex cable   A two-fiber cable suitable for duplex transmission.

Duplex transmission   Transmission in both directions, either one direction at a time (half duplex) or both directions simultaneously (full duplex).

Duty cycle   In a digital transmission, the ratio of high levels to low levels.

Electromagnetic interference   Any electrical or electromagnetic interference that causes undesirable response, degradation, or failure in electronic equipment. Optical fibers neither emit nor receive EMI.

EMD   Equilibrium mode distribution.

EMI   Electromagnetic interference.

Equilibrium mode distribution   The steady modal state of a multimode fiber in which the relative power distribution among modes is independent of fiber length.

Excess loss   In a fiber-optic coupler, the optical loss from that portion of light that does not emerge from the nominally operational ports of the device.

Extrinsic loss   In a fiber interconnection, that portion of loss that is not intrinsic to the fiber but is related to imperfect joining, which may be caused by the connector or splice.

Fall time   The time required for the trailing edge of a pulse to fall from 90% to 10% of its amplitude; the time required for a component to produce such a result. "Turnoff time." Sometimes measured between the 80% and 20% points.

FM Frequency modulation.

Frequency modulation A method of transmission in which the carrier frequency varies in accordance with the signal.

Fresnel reflection The reflection that occurs at the planar junction of two materials having different refractive indices; Fresnel reflection is not a function of the angle of incidence.

Fresnel reflection loss Loss of optical power due to Fresnel reflections.

Gap loss Loss resulting from the end separation of two axially aligned fibers.

Graded-index fiber An optical fiber whose core has a nonuniform index of refraction. The core is composed of concentric rings of glass whose refractive indices decrease from the center axis. The purpose is to reduce modal dispersion and thereby increase fiber bandwidth.

Ground-loop noise Noise that results when equipment is grounded at ground points having different potentials and thereby created an unintended current path. The dielectric of optical fibers provide electrical isolation that eliminates ground loops.

IDP Integrated detector/preamplifier.

Index-matching material A material, used at optical interconnection, having a refractive index close to that of the fiber core and used to reduce Fresnel reflections.

Index of refraction The ratio of the velocity of light in free space to the velocity of light in a given material. Symbolized by n.

Insertion loss The loss of power that results from inserting a component, such as a connector or splice, into a previously continuous path.

Integrated detector/preamplifier A detector package containing a pin photodiode and transimpedance amplifier.

LAN Local area network.

Laser A light source producing, through stimulated emission, coherent, near monochromatic light. Lasers in fiber optics are usually solid-state semiconductor types.

Lateral displacement loss The loss of power that results from lateral displacement from optimum alignment between two fibers or between a fiber and an active device.

LED Light-emitting diode.

Light-emitting diode A semiconductor diode that spontaneously emits light from the pn junction when forward current is applied.

Local area network A geographically limited network interconnecting electronic equipment.

Material dispersion Dispersion resulting from the different velocities of each wavelength in an optical fiber.

MFD Mode field diameter.

Misalignment loss The loss of power resulting from angular misalignment, lateral displacement, and end separation.

Modal dispersion Dispersion resulting from the different transit lengths of different propagating modes in a multimode optical fiber.

Mode In guided-wave propagation, such as through a waveguide or optical fiber, a distribution of electromagnetic energy that satisfies Maxwell's equations and boundary conditions. Loosely, a possible path followed by light rays.

Mode coupling The transfer of energy between modes. In a fiber, mode coupling occurs until EMD is reached.

Mode field diameter The diameter of optical energy in a single-mode fiber. Because the MFD is greater than the core diameter, MFD replaces core diameter as a practical parameter.

Mode filter A device used to remove high-order modes from a fiber and thereby simulate EMD.

Modulation The process by which the characteristic of one wave (the carrier) is modified by another wave (the signal). Examples include amplitude modulation (AM), frequency modulation (FM), and pulse-coded modulation (PCM).

Multimode fiber A type of optical fiber that supports more than one propagating mode.

Multiplexing The process by which two or more signals are transmitted over a single communications channel. Examples include time-division multiplexing and wavelength-division multiplexing.

NA Numerical aperture.

NA-mismatch loss The loss of power at a joint that occurs when the transmitting half has an NA greater than the NA of the receiving half. The loss occurs when coupling light from a source to fiber, from fiber to fiber, or from fiber to detector.

Numerical aperture The "light-gathering ability" of a fiber, defining the maximum angle to the fiber axis at which light will be accepted and propagated through the fiber. $NA = \sin \theta$, where θ is the acceptance angle. NA is also used to describe the angular spread of light from a central axis, as in exiting a fiber, emitting from a source, or entering a detector.

Optical time-domain reflectometry A method of evaluating optical fibers based on detecting backscattered (reflected) light. Used to measure fiber attenuation, evaluate splice and connector joints, and locate faults.

OTDR Optical time-domain reflectometry.

PCM Pulse-coded modulation.

PCS Plastic-clad silica.

Photodetector An optoelectronic transducer, such as a pin photodiode or avalanche photodiode.

Photodiode A semiconductor diode that produces current in response to incident optical power and used as a detector in fiber optics.

Photon A quantum of electromagnetic energy. A "particle" of light.

Pigtail   A short length of fiber permanently attached to a component, such as a source, detector, or coupler.

Pin photodiode   A photodiode having a large intrinsic layer sandwiched between p-type and n-type layers.

Plastic-clad silica fiber   An optical fiber having a glass core and plastic cladding.

Plastic fiber   An optical fiber having a plastic core and plastic cladding.

Plenum   The air space between walls, under structural floors, and above drop ceilings, which can be used to route intrabuilding cabling.

Plenum cable   A cable whose flammability and smoke characteristics allow it to be routed in a plenum area without being enclosed in a conduit.

Pulse-coded modulation   A technique in which an analog signal, such as a voice, is converted into a digital signal by sampling the signal's amplitude and expressing the different amplitudes as a binary number. The sampling rate must be twice the highest frequency in the signal.

Pulse spreading   The dispersion of an optical signal with time as it propagates through an optical fiber.

Quantum efficiency   In a photodiode, the ratio of primary carriers (electron-hole pairs) created to incident photons. A quantum efficiency of 70% means 7 out of 10 incident photons create a carrier.

Rayleigh scattering   The scattering of light that results from small inhomogeneities in material density or composition.

Regenerative repeater   A repeater designed for digital transmission that both amplifies and reshapes the signal.

Repeater   A device that receives, amplifies (and perhaps reshapes), and retransmits a signal. It is used to boost signal levels when the distance between repeaters is so great that the received signal would otherwise be too attenuated to be properly received.

Responsivity   The ratio of a photodetector's electrical output to its optical input in amperes/watt.

Ring network   A network topology in which terminals are connected in a point-to-point serial fashion in an unbroken circular configuration.

Rise time   The time required for the leading edge of a pulse to rise from 10% to 90% of its amplitude; the time required for a component to produce such a result. "Turnon time." Sometimes measured between the 20% and 80% points.

Sensitivity   For a fiber-optic receiver, the minimum optical power required to achieve a specified level of performance, such as a BER.

Shot noise   Noise caused by random current fluctuations arising from the discrete nature of electrons.

Signal-to-noise ratio   The ratio of signal power to noise power.

Simplex cable   A term sometimes used for a single-fiber cable.

Simplex transmission   Transmission in one direction only.

Single-mode fiber   An optical fiber that supports only one mode of light propagation above the cutoff wavelength.

SNR   Signal-to-noise ratio.

Source   The light emitter, either an LED or laser diode, in a fiber-optic link.

Spectral width   A measure of the extent of a spectrum. For a source, the width of wavelengths contained in the output at one half of the wavelength of peak power. Typical spectral widths are 20 to 60 nm for an LED and 2 to 5 nm for a laser diode.

Splice   An interconnection method for joining the ends of two optical fibers in a permanent or semipermanent fashion.

Star coupler   A fiber-optic coupler in which power at any input port is distributed to all output ports.

Star network   A network in which all terminals are connected through a single point, such as a star coupler.

Steady state   Equilibrium mode distribution.

Step-index fiber   An optical fiber, either multimode or single mode, in which the core refractive index is uniform throughout so that a sharp step in refractive index occurs at the core-to-cladding interface. It usually refers to a multimode fiber.

Strength member   That part of a fiber-optic cable composed of Kevlar aramid yarn, steel strands, or fiberglass filaments that increase the tensile strength of the cable.

Tap loss   In a fiber-optic coupler, the ratio of power at the tap port to the power at the input port.

Tap port   In a coupler in which the splitting ratio between output ports is not equal, the output port containing the lesser power.

TDM   Time-division multiplexing.

Tee coupler   A three-port optical coupler.

Thermal noise   Noise resulting from thermally induced random fluctuation in current in the receiver's load resistance.

Throughput loss   In a fiber-optic coupler, the ratio of power at the throughput port to the power at the input port.

Throughput port   In a coupler in which the splitting ratio between output ports is not equal, the output port containing the greater power.

Time-division multiplexing   A transmission technique whereby several low-speed channels are multiplexed into a high-speed channel for transmission. Each low-speed channel is allocated a specific position based on time.

Transducer   A device for converting energy from one form to another, such as optical energy to electrical energy.

Wavelength   The distance between the same two points on adjacent waves; the time required for a wave to complete a single cycle.

Wavelength-division multiplexing   A transmission technique by which separate optical channels, distinguished by wavelength, are multiplexed onto an optical fiber for transmission.

WDM   Wavelength-division multiplexing.

Index